Research Ethics Forum

Volume 8

Series Editors

David Hunter, Flinders University, Adelaide, Australia
John McMillan, University of Otago, North Dunedin, Otago, New Zealand
Charles Weijer, The University of Western Ontario, Ontario, Canada

Editorial Board

Godfrey B. Tangwa, University of Yaounde, Yaounde, Cameroon
Andrew Moore, University of Otago, Otago, New Zealand
Jing-Bao Nie, University of Otago, Otago, New Zealand
Ana Borovečki, University of Zagreb, Zagreb, Croatia
Sarah Edwards, University College London, London, UK
Heike Felzmann, National University of Ireland, Ireland, Ireland
Annette Rid, University of Zurich, Zurich, Switzerland
Mark Sheehan, University of Oxford, Oxford, UK
Robert Levine, Yale University, New Haven, USA
Alex London, Carnegie Mellon University, Pittsburgh, USA
Jonathan Kimmelman, McGill University, Montreal, Canada

This Series, Research Ethics Forum, aims to encourage discussion in the field of research ethics and the ethics of research. Volumes included can range from foundational issues to practical issues in research ethics. No disciplinary lines or borders are drawn and submissions are welcome from all disciplines as well as scholars from around the world. We are particularly interested in texts addressing neglected topics in research ethics, as well as those which challenge common practices and beliefs about research ethics. By means of this Series we aim to contribute to the ever important dialogue concerning the ethics of how research is conducted nationally and internationally. Possible topics include: Research Ethics Committees, Clinical trials, International research ethics regulations, Informed consent, Risk-benefit calculations, Conflicts of interest, Industry-funded research, Exploitation, Qualitative research ethics, Social science research ethics, Ghostwriting, Bias, Animal research, Research participants.

More information about this series at http://www.springer.com/series/10602

M. V. Dougherty

Disguised Academic Plagiarism

A Typology and Case Studies for Researchers and Editors

Springer

M. V. Dougherty
Philosophy Department
Ohio Dominican University
Columbus, OH, USA

ISSN 2212-9529 ISSN 2212-9537 (electronic)
Research Ethics Forum
ISBN 978-3-030-46710-4   ISBN 978-3-030-46711-1 (eBook)
https://doi.org/10.1007/978-3-030-46711-1

© The Author(s), under exclusive license to Springer Nature Switzerland AG 2020
This work is subject to copyright. All rights are solely and exclusively licensed by the Publisher, whether the whole or part of the material is concerned, specifically the rights of translation, reprinting, reuse of illustrations, recitation, broadcasting, reproduction on microfilms or in any other physical way, and transmission or information storage and retrieval, electronic adaptation, computer software, or by similar or dissimilar methodology now known or hereafter developed.
The use of general descriptive names, registered names, trademarks, service marks, etc. in this publication does not imply, even in the absence of a specific statement, that such names are exempt from the relevant protective laws and regulations and therefore free for general use.
The publisher, the authors and the editors are safe to assume that the advice and information in this book are believed to be true and accurate at the date of publication. Neither the publisher nor the authors or the editors give a warranty, expressed or implied, with respect to the material contained herein or for any errors or omissions that may have been made. The publisher remains neutral with regard to jurisdictional claims in published maps and institutional affiliations.

This Springer imprint is published by the registered company Springer Nature Switzerland AG
The registered company address is: Gewerbestrasse 11, 6330 Cham, Switzerland

Acknowledgments

Many people have helped me during the writing of this book. My colleagues at Ohio Dominican University have provided outstanding support, and it is a pleasure to mention Bruce Gartner, Manuel Martinez, Theresa Holleran, Bob Gervasi, and Peter Cimbolic. ODU Philosophy Department members Larry Masek and Brian Besong have given encouragement and critiques. Perry Cahall and Leo Madden provided insightful comments and advice. The library staff at ODU has been very helpful, and I would like to thank Tim Sandusky for his patience and resolve in tracking down many materials.

I have benefited greatly from conversations with Alkuin Schachenmayr, and this book project has taken shape during those valuable discussions. Pernille Harsting has been a wonderful research colleague for over a decade. The research discussed in the latter part of Chap. 4 was conducted with her. I am grateful for recent collaborations with Joshua Hochschild, and much of the background for Chap. 5 was examined with him.

At Springer, I am grateful to editors Floor Oosting and Christopher Wilby for their support for this project. Two reviewers for the press offered detailed and helpful comments on every chapter of the manuscript, and I am thankful for their care in suggesting improvements to the text. Series editors David Hunter, John McMillan, and Charles Weijer kindly included this book in the Research Ethics Forum series.

The Galvin Family Foundation, which established the Sr. Ruth Caspar Chair in Philosophy at ODU, made work on this book possible.

The Spring 2019 honors students enrolled in my Critical Research and Writing course at ODU assisted in examining several of the cases considered in this book. Their insights and critical comments were helpful, and I was privileged to work with them as they begin their own research careers. As a class project, we analyzed a published book chapter for suspected plagiarism and submitted our evidence, with a request for retraction, to the publisher.

In recent years, I have corresponded with journalists who write on academic plagiarism and other violations of research integrity. They include Ivan Oransky, Justin Weinberg, Colleen Flaherty, Alison McCook, Victoria Stern, Shannon Palus,

Adam Marcus, Davide Illarietti, Filipa Silva, and Nikolas Vanhecke. I am grateful for their care in reporting on the principal cases considered in this book.

A portion of the research in this volume was presented at the 2019 Committee on Publication Ethics North American Seminar, and I am grateful to Deborah Poff and the participants for their comments and discussions. I also presented some material for this book at a 2019 research symposium at ODU, and I am appreciative of my faculty colleagues for their valuable questions and comments. This book incorporates revised and updated versions of "The Pernicious Effects of Compression Plagiarism on Scholarly Argumentation," *Argumentation* 33.3 (2019): 391–412 and "The Corruption of Philosophical Communication by Translation Plagiarism," *Theoria* 85.3 (2019): 219–246. I remain thankful to editors Frans van Eemeren and Sven Ove Hansson, and to anonymous reviewers at each journal, for helpful critiques and support.

To my surprise, my children, Thomas, Benedict, and Cecilia have been interested in my work, and I have enjoyed their thoughtful queries and insights. I have discussed all aspects of this book with my wife and colleague, Michelle Dougherty, and I am grateful for her excellent advice, thoughtful criticisms, and good cheer.

Contents

1 **Introduction** 1
   1.1 A Précis of Chapters 4
   1.2 How Plagiarism Corrupts the Published Research Literature 6
   1.3 In Sum 9
   References 11

2 **Translation Plagiarism** 13
   2.1 Defining Translation Plagiarism 14
   2.2 The Prevalence of Translation Plagiarism 14
   2.3 A Case of Translation Plagiarism 15
   2.4 Overlap with Peter Stemmer's *Platons Dialektik* 16
      2.4.1 The Role of Disguising References 18
      2.4.2 The Hidden German Phase of N. 2001 19
      2.4.3 The Hypothesized Document of the Hidden German
            Phase 20
      2.4.4 Relay Translation Plagiarism 22
      2.4.5 Duplication in the Downstream Literature 23
      2.4.6 Corruptions of the Downstream Literature 24
   2.5 Overlap with Stefan Gosepath's *Aufgeklärtes Eigeninteresse* 25
      2.5.1 Meta-narrative in N. 2001 26
      2.5.2 The Repetition of Idiosyncratic Examples 28
      2.5.3 Bibliographical Overlap 29
      2.5.4 Strong and Weak Rationality 30
      2.5.5 A Shared Thesis 30
      2.5.6 The Downstream Literature Problem 32
   2.6 Other European Languages 32
   2.7 Conclusion 33
   2.8 Postscript 34
   References 34

3 Compression Plagiarism ... 37
3.1 A Case of Compression Plagiarism ... 38
3.2 How Compression Plagiarism Corrupts Scholarly Communication ... 41
3.3 Objections ... 44
3.4 Conclusion ... 46
3.5 Postscript ... 47
Appendix ... 47
References ... 49

4 Dispersal Plagiarism ... 51
4.1 Case 1: The Clandestine Afterlife of a 2003 Dissertation ... 53
    4.1.1 The Role of Electronic Thesis Depositories ... 55
    4.1.2 Different Titles and Competing Copyright Claimants ... 56
    4.1.3 Instances of Plagiarism from Now-Retracted Articles ... 57
    4.1.4 The 2003 Kribbe Thesis in the Body of Published Research Literature ... 61
4.2 Case 2: The Post-publication Re-Emergence of a 1994 Monograph ... 62
    4.2.1 Examples of Plagiarism ... 64
    4.2.2 Kantola Versus S. in the Downstream Literature ... 64
4.3 Case 3: The English Re-Incarnation of a German Monograph ... 67
    4.3.1 Dispersal Plagiarism in Translation ... 69
4.4 Conclusion ... 71
References ... 72

5 Magisterial Plagiarism ... 75
5.1 Cardinal William Levada's Ghostwriter ... 76
    5.1.1 A Subtle Change in Subject Matter ... 77
    5.1.2 First-Person Plagiarism ... 79
5.2 Cardinal Marc Ouellet's Ghostwriter ... 81
    5.2.1 Double Plagiarism ... 81
    5.2.2 Other Sources Appropriated by Ouellet's Ghostwriter ... 85
5.3 A Second Production by Ouellet's Plagiarizing Ghostwriter ... 86
5.4 A Third Production by Ouellet's Plagiarizing Ghostwriter ... 89
5.5 The Downgrading of Magisterial Texts ... 91
5.6 Who Is the Plagiarizing Ghostwriter for Cardinal Ouellet? ... 93
    5.6.1 The Circumstantial Evidence ... 94
    5.6.2 The Textual Evidence ... 94
    5.6.3 Triple Plagiarism ... 96
5.7 Conclusion ... 98
References ... 99

6	**Exposition Plagiarism**		103
	6.1 Entirely Unattributed Texts		105
	6.2 Deficient Attribution		108
	6.3 Incorrect Attribution		110
	6.4 Presenting Debates in the Secondary Literature		111
	6.5 Primary Texts in Philosophy		116
	6.6 Primary Texts in Biology		118
	6.7 In Sum		122
	6.8 Postscript		123
	References		125
7	**Template Plagiarism**		127
	7.1 Standard Template Plagiarism		128
		7.1.1 A Change of Region	129
		7.1.2 A Change in Topic	130
		7.1.3 Changes in Region and Topic	132
		7.1.4 A Change in Genre	133
	7.2 Template Plagiarism Versus Template Text Recycling		134
	7.3 Template Plagiarism Involving Data Fabrication		135
		7.3.1 Fieldwork Interviews and Template Plagiarism	136
		7.3.2 Evaluating Evidence of Template Plagiarism Involving Fieldwork	139
	7.4 The Misuse of Anonymity Protections with Template Plagiarism		143
		7.4.1 Triplicate Publication and Template Plagiarism	143
	7.5 Sources for Template Plagiarism		145
	7.6 A Question of Scale		146
	7.7 Conclusion		147
	References		148
8	**Conclusion: Remedies for Disguised Plagiarism**		151
	8.1 Resources and Tools		152
	References		153
Index			155

About the Author

M. V. Dougherty holds the Sr. Ruth Caspar Chair in Philosophy at Ohio Dominican University (USA). He is author of *Correcting the Scholarly Record for Research Integrity: In the Aftermath of Plagiarism* (Springer, 2018) and *Moral Dilemmas in Medieval Thought: From Gratian to Aquinas* (Cambridge University Press, 2011). He has edited *Aquinas's Disputed Questions on Evil: A Critical Guide* (Cambridge University Press, 2016) and *Pico della Mirandola: New Essays* (Cambridge University Press, 2008). His research interests include research ethics and the history of ethics. For over a decade, he has been involved in securing dozens of retractions, errata, and corrigenda for published articles in the discipline of philosophy and in related fields. His work in generating corrections for academic plagiarism and various authorship violations has been featured on *Retraction Watch* and on other academic news outlets.

List of Figures

Fig. 1.1	Additions to the published research literature by responding articles	7
Fig. 1.2	Corruption of the downstream research literature by plagiarism	8
Fig. 1.3	The whistleblower as undeceived reader	10
Fig. 2.1	How typical readers encounter N. 2001's account of Herodotus	19
Fig. 2.2	Disguised textual transmission in N. 2001	20
Fig. 2.3	N. 2001 as a proxy for Stemmer 1992	23
Fig. 4.1	Distribution of bibliographical references in A. 2011, 2012	60
Fig. 4.2	The journey of two sentences from Kribbe 2003	62
Fig. 5.1	The textual transmission of Driscoll's text	83
Fig. 5.2	The substitution of key terms by two plagiarizing ghostwriters	85
Fig. 5.3	Driscoll's text in five plagiarizing compilations	98
Fig. 6.1	How the interpretation of G. 2018a appears to a reader	120
Fig. 6.2	The hidden transmission of Oken's text via Turk's translation	120

List of Tables

Table 2.1	N. 2001 and Stemmer 1992 on the fifth-century crisis period/Krisenzeit	17
Table 2.2	A selection of Herodotus translations	18
Table 2.3	N. 2003 as the German text for N. 2001, with Stemmer 1992 in bold	21
Table 2.4	N. 2001 and Gosepath 1992 on whether an opinion/Meinung is rational/rational	26
Table 2.5	N. 2001 and Gosepath 1992 on the conditions of rationality	27
Table 2.6	Examples common to N. 2001 and Gosepath 1992	28
Table 2.7	Parallel bibliographical sources	29
Table 2.8	Shared key terms in N. 2001 and Gosepath 1992	30
Table 2.9	Rationality as "well-founded"/"wohlbegründet" in N. 2001 and Gosepath 1992	31
Table 2.10	English-to-French textual overlap	33
Table 3.1	A display of passages from N. 2001 and N. 2006 in relation to Gosepath 1992	45
Table 3.2	Compression (a)	48
Table 3.3	Compression (b)	48
Table 3.4	Compression (c)	48
Table 3.5	Compression (d)	49
Table 3.6	Compression (e)	49
Table 3.7	Compression (f)	49
Table 3.8	Compression (g)	49
Table 4.1	Overlap between A. 2008a–A. 2012 and Kribbe 2003	54
Table 4.2	Summary of overlap between A. 2008a–A. 2012 and Kribbe 2003	55
Table 4.3	Triplicate publication of a source text	57
Table 4.4	Duplicate publication of a source text	58
Table 4.5	Plagiarism without duplicate publication	60
Table 4.6	Overlap between S. et al. 1999–S. 2005 and Kantola 1994	63

Table 4.7	Summary of overlap between S. et al. 1999–S. 2005 and Kantola 1994	63
Table 4.8	Kantola's account of Suárez in three works in triplicate publication by S	65
Table 4.9	Kantola's account of Henry of Ghent in triplicate publication by S	66
Table 4.10	Overlap between N. 2002–N. 2014 and Wieland 1982	68
Table 4.11	Summary of overlap between N. 2002–N. 2014 and Wieland 1982	69
Table 4.12	Translation plagiarism and text recycling on propositional statements	70
Table 4.13	Translation plagiarism and text recycling on the content and structure of sentences	70
Table 4.14	Translation plagiarism and duplicate publication on Socratic dialectic	71
Table 5.1	A change in subject matter from Driscoll 2000/2002/2003 to Levada 2007	77
Table 5.2	From preaching to theology	78
Table 5.3	Driscoll on the function of Patristic preaching	79
Table 5.4	Plagiarism and the use of the first person	80
Table 5.5	The presence of Driscoll's text in Levada 2007 and Ouellet 2007	82
Table 5.6	The institutional hopes of two cardinals	85
Table 5.7	Ouellet's ghostwriter's use of Schillebeeckx	86
Table 5.8	Ratzinger 2005 in Ouellet 2008a	88
Table 5.9	A VIS brief and Ouellet 2008a	88
Table 5.10	Writings by John Paul II in Ouellet 2008b	89
Table 5.11	The patchwork plagiarism style of Cardinal Ouellet's ghostwriter	91
Table 5.12	Allen 2004 and Ouellet 2007	94
Table 5.13	R. 2006 and Ouellet 2007 on John Paul II	96
Table 5.14	Ouellet 2007 as an undisclosed source text for R. 2011b	97
Table 6.1	Sunderland 2015 and an undisclosed source text for G. 2018a	106
Table 6.2	G. 2018a and Roe 1981 on C. F. Wolff's notion of essential force	107
Table 6.3	Dupont 2018's abstract and G. 2018a	109
Table 6.4	Incorrect attribution and no quotation marks	110
Table 6.5	Ginsborg's summary of her position as repeated in G. 2018a	112
Table 6.6	Lenoir's summary of his position as presented in G. 2014	113
Table 6.7	Lenoir's summary of his position as presented in G. 2018a	114

Table 6.8	Selections of Peterson 2004 in G. 2014 and G. 2018a	115
Table 6.9	Kant's views on Kant in G. 2018a	116
Table 6.10	Hegel's views on Hegel in G. 2018a	117
Table 6.11	Smith 2011 on Leibniz's views on animal bodies in G. 2018a	118
Table 6.12	Oken's view on Oken in G. 2018a	119
Table 6.13	Haller's views on Haller in G. 2018a	122
Table 7.1	The substitution of a region for a continent	129
Table 7.2	The substitution of a subject term	130
Table 7.3	A change in references to secondary literature	131
Table 7.4	A change in example countries	131
Table 7.5	A change in country and in topic	132
Table 7.6	A template in passages on Poland, Croatia, and Albania	134
Table 7.7	A Macedonian fieldwork interview statement concerning CSO funding	137
Table 7.8	A Macedonian fieldwork interview statement on social progress	138
Table 7.9	A Serbian fieldwork interview on anti-corruption laws	139
Table 7.10	A Serbian fieldwork interview on criminal influences	140
Table 7.11	A Serbian fieldwork interview on law enforcement	141
Table 7.12	Suspected template plagiarism versus suspected copy-and-paste plagiarism	141
Table 7.13	Three versions of a passage alongside its suspected source text	144
Table 7.14	Template plagiarism in economic history	146

Chapter 1
Introduction

Abstract This chapter introduces a book-length study of under-recognized forms of disguised plagiarism that mar humanities disciplines. All forms of disguised plagiarism involve some additional concealment beyond the mere copying of words with inadequate attribution. The various kinds of secondary concealment create a further distance between the source text and the plagiarizing text that renders the plagiarism more difficult to identify. This book defends a typology of disguised plagiarism consisting of six principal forms. Using cases from the disciplines of philosophy, history, theology, and political science, I argue that the proposed typology is useful for understanding the precise ways that plagiarizing books and articles contaminate the body of published research literature in humanities disciplines. Since academic plagiarists in humanities often combine forms of disguised plagiarism, a precise evaluation of the inauthenticity of a given work presupposes an understanding of the distinct kinds of disguised plagiarism.

Keywords Disguised plagiarism · Humanities · Research ethics · Research integrity · Publication ethics

Over the last decade I have encountered variations of the claim "plagiarism is easy to detect" in discussions about publishing integrity and scholarly writing (e.g., Valiela 2009: 288; Wallwork 2016: 187). My experience has been otherwise. With the help of colleagues, I have requested retractions for more than 125 plagiarizing articles in humanities fields in recent years. A large portion of these articles exhibited very subtle forms of plagiarism. Many researchers, editors, and publishers in humanities appear to be unaware of the disguised varieties of plagiarism and the resultant damage that subtly plagiarizing articles inflict on the body of published research. When undetected, plagiarizing articles produce widespread inefficiencies in the wider system of knowledge production. Not only are researchers denied credit for their discoveries, but plagiarizing articles take up space in journals that should have been reserved for articles by authentic researchers. On the basis of these fraudulent publications, successful plagiarists take up other already scarce resources in humanities fields, including academic teaching and research posts, grants, promotions, and awards. In some instances, successful plagiarists have gone on to serve as editors and board members of academic journals, and they have edited book series with well-established publishers.

The best defense against these corruptions is a recognition of the varieties of plagiarism. *Plagiarism* is not a univocal term; to conceive of it exclusively in terms of the straightforward copy-and-paste variety is an inadequate approach to research misconduct. My purpose in this book is to set forth in clear terms the varieties of disguised plagiarism, so that researchers, editors, and publishers can become familiar with the subtler forms of plagiarism that corrupt the production and dissemination of knowledge in humanities fields. I intend the volume to be useful and have supplemented the theoretical discussions of plagiarism with clear cases studies, examples, and flow charts to assist researchers, editors, and publishers. The book provides a concise etiology of the problem of disguised plagiarism. A greater awareness of the varieties of disguised plagiarism will support a culture of vigilance. Left unchecked, disguised plagiarism will continue to threaten the reliability of the enterprise of research.

Much excellent work has been done in recent years to classify the more straightforward varieties of plagiarism (Weber-Wulff 2014: 6–14; Zhang 2016: 9–10). One useful approach divides plagiarism into the major classes of *literal* and *disguised* (Gipp 2014: 11–13). Literal plagiarism consists of those forms that are the easiest to detect, such as copy-and-paste plagiarism, which involves the verbatim or near-verbatim appropriation of a source text. Another form of literal plagiarism, shake-and-paste plagiarism, is defined as "copying and merging of text segments with slight adjustments to form a coherent text, e.g., by changing word order, by substituting words with synonyms, or by adding or deleting 'filler' words" (Gipp 2014: 11; see Weber-Wulff 2014: 8–9). A third form, pawn-sacrifice plagiarism, occurs when "the source citation is either given in a footnote or only listed in the bibliography" and the plagiarist has "not made clear, however, exactly how much has been taken" (Weber-Wulff 2014: 10; see Lahusen 2006).

Most of the work on plagiarism typologies has focused on literal plagiarism, which is more likely to be discovered by text-matching software (and by careful readers) than disguised plagiarism. Disguised plagiarism is more subtle; all its varieties involve some additional concealment that creates further distance between the plagiarizing text and its source. These forms are not only the most difficult to detect, but also are the most challenging to demonstrate. Persuading fellow researchers, editors, and publishers of cases of them is often not easy.

In this book, I retain the helpful concept of *disguised plagiarism*, but with a major caveat. Some writers on research integrity include a wide variety of research integrity violations under the heading of "plagiarism." They include such violations commonly denominated as "idea plagiarism," "structural plagiarism," and "self-plagiarism" (Gipp 2014: 12–13). Furthermore, some researchers include the referencing of non-existing sources as a form of plagiarism (e.g., Tauginienė et al. 2019: 11, 2018: 26). In contrast to these approaches, I propose a taxonomy of disguised plagiarism that is restricted to situations involving the republication of the words of a source text under new authorship. The theft of ideas ("idea plagiarism"), the re-purposing of the structure of another's argument without credit ("structural plagiarism"), the undisclosed republication of one's own work ("self-plagiarism"), and the manufacture of fictitious citations ("invalid source plagiarism") are each serious violations of good research practices. Nevertheless, classifying these violations under heading of *plagiarism* weakens the utility of the concept of plagiarism as a precise category of research misconduct. In this book, I restrict the term *plagiarism*—in both its disguised and undisguised varieties—to those forms of plagiarism involving the verbatim and near-verbatim re-use of another's words with inadequate credit.

This book presents six under-recognized forms of disguised plagiarism that corrupt the quality of published research in humanities disciplines. By the expression *disguised plagiarism*, I mean plagiarism that exhibits some additional form of concealment beyond the mere removal of the original author's name from a text and presenting it as one's own. The chapters that follow identify and examine the following phenomena:

- **Translation plagiarism**: changing a source text into a different language
- **Compression plagiarism**: distilling a lengthy scholarly text into a short one
- **Dispersal plagiarism**: dividing a text and publishing it in several venues
- **Magisterial plagiarism**: concealing the authoritative status of ecclesiastical texts in theological writing
- **Exposition plagiarism**: conflating the authorial voices of canonical authors, of commentators, and of the plagiarist in historiographical writing
- **Template plagiarism**: replacing the key terms of a source text to produce the illusion of new research on a different topic.

I provide case studies to illustrate each type of disguised plagiarism and explore the particular harm that the various kinds of disguised plagiarism inflict on disciplines in humanities. As shall be seen below, I am particularly interested in cases where a single plagiarizing book or article exhibits more than one form of disguised plagiarism.

In the analyses that follow, readers will not find the names of confirmed or suspected plagiarists. My concern is to identify and catalogue forms of disguised plagiarism, not to study plagiarists themselves. My research focus is strictly to present a clear and useful typology, rather than to investigate persons who plagiarize; I have no interest in speculating on their motivations or directing further attention to them for any of their academic misdeeds. (In most cases, their acts of plagiarism have received news coverage in academic outlets in North America and in Europe.) For this reason, I refer to each of them using an abbreviation ("N."; "E."; "S."; "A."; "R."; "G.", "C.", "F."). In two cases where the precise identity of the plagiarist is unknown, I refer to the plagiarist by a position description (e.g., "the ghostwriter for *x*"). Nearly all of these authors of record for defective articles have already received published corrections by editors and publishers for failures to acknowledge source texts adequately. These corrections have taken the form of retractions, corrigenda, or errata, where the publisher authoritatively changes the version of record of the defective article. These corrections are the result of third-party reporting to editors and publishers. (This fact underscores the importance of all members of the academic community—not just publishers and editors—in working vigilantly to maintain the reliability of the body of published research literature.) Most authors of record of these corrected articles have received multiple retractions, and some have reached double-digit numbers. High numbers of retractions for plagiarists should not be too surprising, since academic plagiarism is almost always serial (Fox and Beall 2014: 346).

A retraction is an official designation by a publisher that a work is now regarded as substantively untrustworthy, and for this reason retractions have been described as the "nuclear option" for editors and publishers for dealing with research misconduct (Marcus and Oransky 2017: 119). Retraction is the chief post-publication remedy for minimizing the harm of a plagiarizing article or book. Even among new proposals for expanding of the kinds of corrections available to editors and publishers, the role of retractions for cases of fraud and proven research misconduct remains unchanged (Fanelli et al. 2018).

The detailed analyses of plagiarizing books and articles in the chapters that follow illustrate how to demonstrate plagiarism. It is one thing to believe that an article is defective, and quite another thing to identify and display evidence of plagiarism in a way that is persuasive, factual, and publicly verifiable. *Proving plagiarism involves building a case. Knowing the different kinds of evidence, and knowing how to present them, is crucial to success.* This book can be used as a guide for whistleblowers for proving disguised forms of plagiarism as well as a guide for editors and publishers for evaluating submitted evidence of disguised plagiarism.

1.1 A Précis of Chapters

This book is divided into eight chapters. In the next chapter, I consider an especially subtle type of disguised plagiarism, *translation plagiarism*, which occurs when the work of one author is republished in a different language with authorship credit taken by someone else. I focus on the challenges of demonstrating this variety of plagiarism and examine the corruptive influence that such plagiarizing articles exert on unsuspecting researchers who later cite them in the downstream literature as genuine products of research. Although translation plagiarism has received some attention in the field of translation studies, little attention has been given to the pernicious effects of translation plagiarism in published books and articles of academic philosophy. This omission should not be surprising; identifying cases of translation plagiarism is difficult, and presenting evidence can be challenging. Furthermore, receptive venues for considering the problem are few. Nevertheless, open discussions of translation plagiarism can contribute to improving the integrity of scholarship. I conclude by arguing that an open discussion of plagiarizing articles is necessary for maintaining the reliability of the body of published research and for restoring integrity to scholarly communication.

In the third chapter, I provide an analysis of one under-recognized disguised variety of plagiarism—designated as *compression plagiarism*—that consists of the distillation of a lengthy scholarly text into a short one, followed by the publication of the short one under a new name with inadequate credit to the original author. In

1.1 A Précis of Chapters

typical cases, a book is compressed into an article, or a long article is compressed into a very short one. Compression plagiarism is a form of disguised plagiarism because the compression of the source text further obscures the plagiarizing text's origin. The compression may be twofold. First, the compression may occur at the macro level and involve the appropriation of passages that are very distant in the source text. For example, a plagiarist might take portions from the introduction, central chapters, and conclusions of a lengthy book in producing a short journal article. Second, the compression may occur at the micro level and involve the reduction of a paragraph into a sentence (or a long sentence into a shorter one). Compression of both varieties impedes a typical reader from recognizing the dependency of the plagiarizing article on its lengthy source text. I argue that this kind of plagiarism is especially damaging to the quality of scholarly argumentation.

The fourth chapter identifies the phenomenon of *dispersal plagiarism* and analyzes its negative effects. Dispersal plagiarism occurs when a lengthy single work by a genuine author is divided by a plagiarist who then publishes portions of it in various venues without credit to the genuine author. A typical case involves the appropriation of a book by a plagiarist who publishes the chapters (or sections of chapters) in discrete articles across several journals. Subsequent researchers are more likely to encounter the scholarly contributions of the genuine author not in its original book format but in one or more of the plagiarizing versions. In the downstream literature, citations accrue to the several plagiarizing versions instead of to the original, and the body of published literature is thereby corrupted. If dispersal plagiarism is combined with duplicate publication and text recycling, the original author's work can virtually disappear from the research literature as citations migrate to the more recent and numerous plagiarizing versions. To exhibit the negative impact of dispersal plagiarism, I consider three distinct cases of the phenomenon from the published research literature in the discipline of philosophy.

The fifth chapter considers an affirmative answer to the question of whether plagiarism can cause more harm when perpetrated in some academic disciplines than in others. I examine the effects of plagiarism committed in the specific disciplinary context of theology, and in particular, Catholic theology. Members of the Catholic hierarchy regularly produce texts considered to be authoritative for the faithful. These magisterial texts, authored by popes, cardinals, and bishops, shape the contours of Catholic theology in its various subfields, including work done in systematic, biblical, and historical theology. *Magisterial plagiarism* occurs in two ways. First, when ghostwriters for members of the Catholic hierarchy secretly turn in plagiarizing work to their unsuspecting employers, the subsequent promulgation of these defective texts by members of the Catholic hierarchy harms the quality of magisterial teaching that informs the practice of theology. Second, when theologians themselves plagiarize magisterial texts in their own theological writings, falsely presenting the words of popes, cardinals, and bishops as their own, the prior magisterial endorsement of the source texts is concealed to readers. The magisterial authority of these texts is disguised or masked in the act of plagiarism. Unsuspecting readers are thereby disadvantaged in their attempts to interpret the texts and assess their weight.

The sixth chapter considers forms of plagiarism in historiographical writing. Plagiarizing historiographical writing generates problems for readers, not the least of which is confusion concerning authority. In a well-written, non-plagiarizing study, readers are presented with a clear separation of three kinds of authority: (1) the voices of authors of primary canonical texts whose works are cited; (2) the voices of exegetes in the secondary literature whose interpretations are analyzed, and (3) the voice of the author of the well-written work that contains an innovative account. In contrast, plagiarizing historiographical studies fail to demarcate clearly these three kinds of authority, and this failure vitiates the quality of that work. The chapter explores the phenomenon of *exposition plagiarism*, understood as plagiarism that involves the conflating of authoritative voices in historiographical writing.

The penultimate chapter examines the phenomenon of *template plagiarism*, which occurs when a plagiarist uses a previously published passage on one subject and reworks it to produce a seemingly new passage on a different subject by changing a key term. For example, altering a passage that discusses one country by substituting the name of another country produces the illusion of original research on the second country. Most other forms of plagiarism—disguised as well as undisguised—require that the source text and the plagiarizing text concern the same subject matter. This limitation does not generally apply, however, to cases of template plagiarism. The substitution of key terms allows a plagiarist to generate the appearance of novel research on a topic that is unrelated to the source text. In its more complex variety, template plagiarism may be co-extensive with another form of research misconduct—data fabrication—when the source text involves quantitative or qualitative research.

A brief final chapter completes the volume. It identifies resources that support editors, publishers, and whistleblowers in maintaining the integrity of the body of published research.

1.2 How Plagiarism Corrupts the Published Research Literature

Plagiarism theorist Debora Weber-Wulff rightly observes that "plagiarism is a continuous spectrum of text manipulations and not just one particular method of using other people's words" (2014: 14). If the varieties of plagiarism are neither clearly understood nor clearly defined, then some defects in the published research literature will likely remain uncorrected. A conceptual analysis of the varieties of plagiarism can assist researchers—as well as editors and publishers—in identifying cases of defective scholarship. Such conceptual analysis can also assist editors and publishers in formulating fully explanatory statements of retraction to restore the reliability of the set of published works for each field.

The way plagiarizing articles mar the reliability of the body of published research articles is not always well understood. As soon as any plagiarizing article is issued by a publisher in its final version of record, it can be cited by other researchers in

1.2 How Plagiarism Corrupts the Published Research Literature

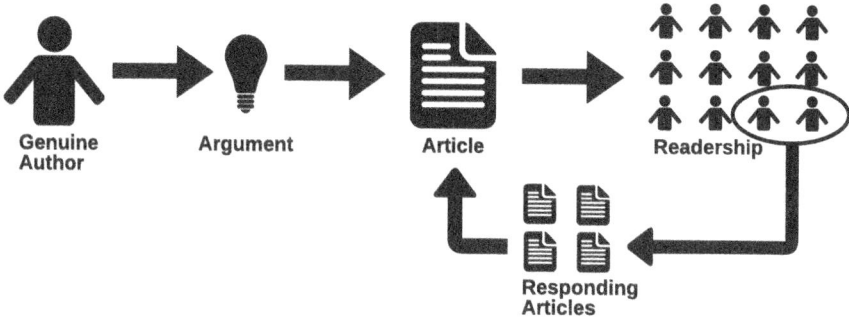

Fig. 1.1 Additions to the published research literature by responding articles

their own articles. Any delay in the issuance of a retraction for plagiarism increases the time in which unsuspecting researchers will mistake the plagiarizing article to be a genuine product of research. These unsuspecting researchers then unwittingly cite the plagiarizing article as trustworthy in their own research articles, further embedding the plagiarizing article in the published body of research. A little-observed consequence of unretracted plagiarizing articles is the damage they inflict when innocent researchers unwittingly incorporate the defective work in their own research. Unless timely retractions are issued for plagiarizing articles, more readers of defective works will become the unsuspecting maleficiaries of the work of plagiarists.

To illustrate this phenomenon, one may compare the case where a good-faith researcher cites a legitimate product of research against the case where a good-faith researcher unknowingly cites a plagiarizing article. The first scenario is illustrated in Fig. 1.1, which represents the typical way the body of published research is increased in a given field.

In cases such as this one, a genuine author discovers an argument that is then published as an article, and the readership of the article eventually includes those who themselves will produce their own responding articles that engage the first article. A field expands its body of truth claims when practitioners conduct their research in light of what came before and then display their findings in subsequent responding articles.

A typical act of plagiarism, however, generates a downstream corruption after a genuine article has been added to the body of published research literature. Figure 1.2 illustrates how an act of plagiarism and subsequent responses to a plagiarizing article damage the reliability of the body of published research.

The readership of a genuine article sometimes includes a plagiarist who repurposes the argument of the article into a new article, with insufficient attribution, in the manufacture of a plagiarizing article. The subsequent readership of this plagiarizing article is then deceived about the origin of the argument it contains. When members of the deceived readership later produce their own articles that engage the plagiarizing article, they unwittingly generate corrupted responding articles that enter the body of published research literature.

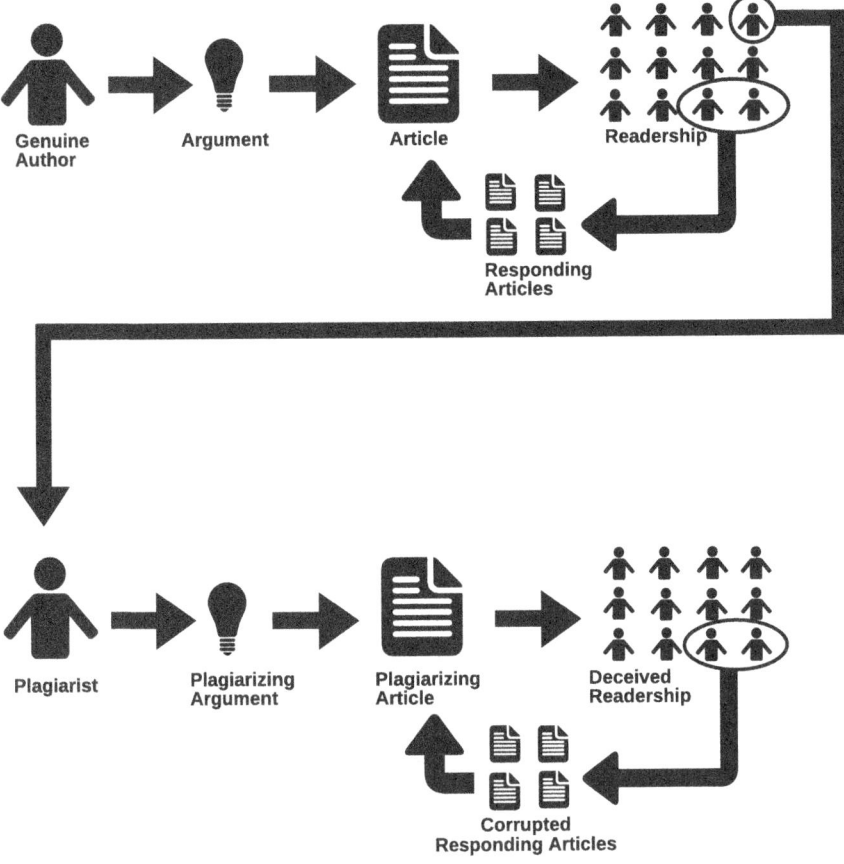

Fig. 1.2 Corruption of the downstream research literature by plagiarism

These responding articles are corrupted insofar as they blindly engage the argument of the original author through the proxy of the plagiarist. The argument originally produced by the genuine author has been promulgated *twice* within the body of published research: first in the original article and then in the plagiarizing article. If the plagiarist then engages in duplicate or redundant publication (what is sometimes called "self-plagiarism"), by republishing the plagiarizing text in another venue, the cycle is repeated.

Responding articles are thus divided in the published research literature. Some respond to the argument as openly expressed in the genuine author's original article, and some respond to the argument as concealed in the plagiarizing article. The plagiarizing article and corrupted responding articles each obscure the origin of the argument in the genuine author's article. The genuine author is denied credit for her or his original discoveries, and the plagiarist receives undeserved laurels.

1.2 How Plagiarism Corrupts the Published Research Literature

The twofold presence of the genuine author's argument both in the original article and in the plagiarizing article creates a redundancy in the published research literature. This duplication generates inefficiencies for researchers working through the research literature in a given field. The redundancy is further magnified because there are also two sets of responding articles to the same argument: (1) those that respond to the genuine author's original article, and (2) those in response to the plagiarist's article.[1]

When editors and publishers issue official retractions of plagiarizing articles without delay, they minimize the chances of continued citations to the plagiarizing articles, thereby stemming the production of additional corrupted responding articles. The issuance of a retraction by an editor or publisher is typically the culmination of a process initiated by a third-party whistleblower. A plagiarism whistleblower typically belongs to the original readership of the genuine article and then later encounters the plagiarizing article but is not deceived by it. Figure 1.3 illustrates the whistleblower's relationship to the genuine article and the plagiarizing article.

Acts of whistleblowing are essential to maintaining the integrity of the body of published research in the aftermath of plagiarism (Fox and Beall 2014; Dougherty 2018: 117–143). Retractions correct the published research literature and restore its reliability, particularly when retractions explicitly give credit to the original authors whose works have been misappropriated by plagiarists and are issued in a way that is easily accessible to the research community (Bilbrey et al. 2014; Dougherty 2018: 95–97; Vuong 2020).

1.3 In Sum

Disguised plagiarism is no less damaging to the integrity of the published research literature than blatant copy-and-paste plagiarism. Subtle forms of plagiarism are simply more difficult to identify. In all its varieties, plagiarism creates the false impression that the author of record is the genuine author of what appears in print; the original, victimized author remains obscured to readers. The vast majority of readers of a plagiarizing article have no way of determining, at least from the text itself, that the author of record is not the genuine author. When a plagiarist forgoes traditional conventions such as the use of quotation marks, in-text citations, extract quoting, and the like in manufacturing a plagiarizing article, the reader is typically unable to determine the parts that are original from the parts that are not. A reader's default assumption is likely to be that the author of record, whose name appears on the published work, is the originator of everything that is not explicitly credited as originating elsewhere.

---

[1] Inefficiencies and redundancies are created in the body of published research literature by responding articles in other ways also. For the phenomena of "citation amnesia," "cryptomnesia," or "disregard syndrome," see Hansson (2008: 99); Ginsburg (2001: 51).

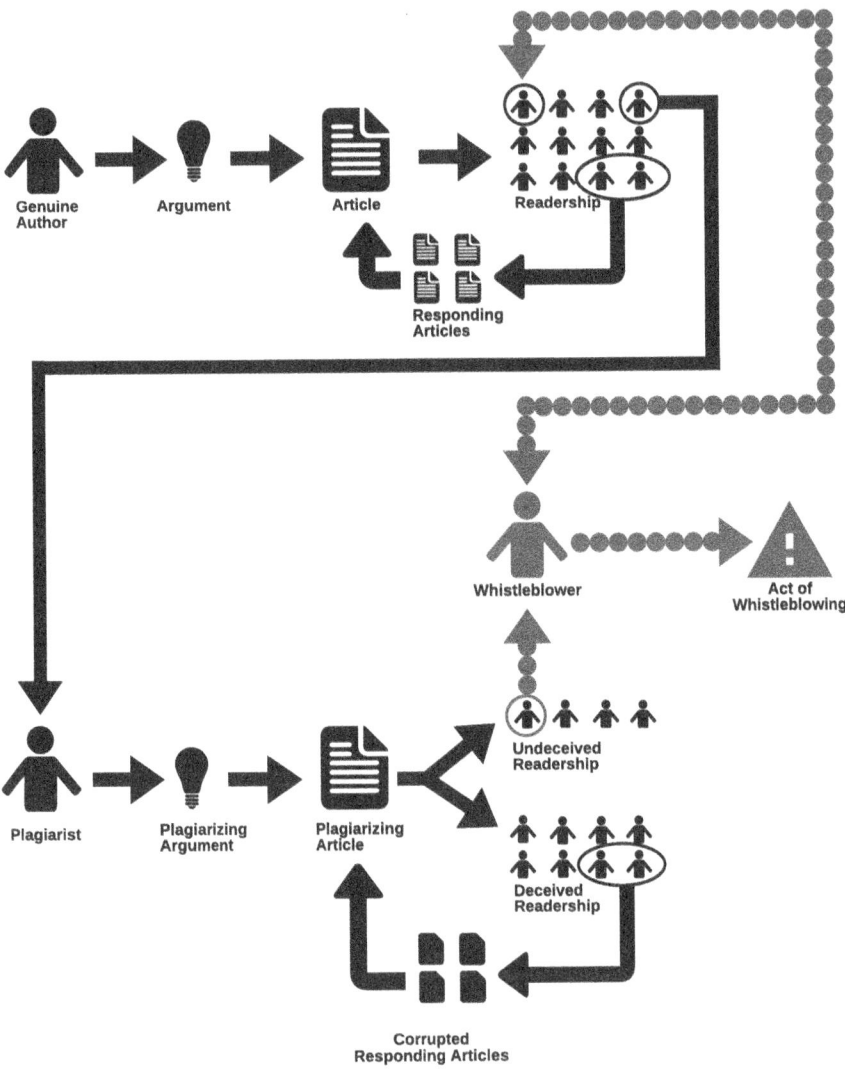

Fig. 1.3 The whistleblower as undeceived reader

The main purpose of this book is to set forth a useful typology of disguised plagiarism varieties. The typology set forth here is supplemented by detailed case studies that demonstrate the real existence of each form of disguised plagiarism. Without a sufficiently refined typology of disguised plagiarism, reliable determinations regarding the authenticity of books and articles cannot be made. Furthermore, since some successful serial plagiarists combine forms of disguised plagiarism in a single plagiarizing book or article, an accurate diagnosis of the defects presupposes a sufficiently sophisticated typology. If plagiarizing articles cannot be diagnosed properly, they will not be identified as defective by researchers or retracted by editors and publishers. Their enduring presence in the body of published research will continue to pollute the downstream literature as researchers read them under the assumption that they are authentic products of research.

References

Bilbrey, Emma, Natalie O'Dell, and Jonathan Creamer. 2014. A novel rubric for rating the quality of retraction notices. *Publications* 2 (1): 14–26. https://doi.org/10.3390/publications2010014.
Dougherty, M.V. 2018. *Correcting the scholarly record for research integrity*. Cham: Springer.
Fanelli, Daniele, John P. A. Ioannidis, Steven Goodman. 2018. Improving the integrity of published science. *European Journal of Clinical Investigation* 48(4): 1–6, e12898.
Fox, Mark, and Jeffrey Beall. 2014. Advice for plagiarism whistleblowers. *Ethics and Behavior* 24 (5): 341–349.
Ginsburg, Isaac. 2001. The disregard syndrome. *The Scientist* 15 (24): 51.
Gipp, Bela. 2014. *Citation-based plagiarism detection*. Wiesbaden: Springer Vieweg.
Hansson, Sven Ove. 2008. Philosophical plagiarism. *Theoria* 74 (2): 97–101.
Lahusen, Benjamin. 2006. Goldene Zeiten. *Kritische Justiz* 39 (4): 398–417.
Marcus, Adam, and Ivan Oransky. 2017. Is there a retraction problem? In *The Oxford handbook of the science of science communication*, ed. Kathleen Hall Jamieson, et al., 119–126. New York: Oxford University Press.
Tauginienė, Loreta, et al. 2018. *Glossary for academic integrity*. ENAI. https://www.academicintegrity.eu/wp/wp-content/uploads/2018/02/GLOSSARY_final.pdf.
Tauginienė, Loreta, et al. 2019. Enhancing the taxonomies relating to academic integrity and misconduct. *Journal of Academic Ethics* 17 (4): 345–361.
Valiela, Ivan. 2009. *Doing science*, 2nd ed. Oxford: Oxford University Press.
Vuong, Quan-Hoang. 2020. The limitations of retraction notices. *Learned Publishing* 33 (2): 119–130.
Wallwork, Adrian. 2016. *English for writing research papers*. Cham: Springer.
Weber-Wulff, Debora. 2014. *False feathers*. Heidelberg: Springer.
Zhang, Yuehong (Helen). 2016. *Against plagiarism*. Cham: Springer.

Chapter 2
Translation Plagiarism

Abstract Disguised plagiarism often goes undetected. An especially subtle type of disguised plagiarism is *translation plagiarism*, which occurs when the work of one author is republished in a different language with authorship credit taken by someone else. I focus on the challenge of demonstrating this subtle variety of plagiarism and examine the corruptive influence that plagiarizing articles exert on unsuspecting researchers who later cite them in the downstream literature as genuine products of research. I conclude by arguing that an open discussion of plagiarizing articles in philosophy is necessary for maintaining the reliability of the body of published research and for restoring integrity to scholarly communication.

Keywords Philosophical communication · Plagiarism · Authorship · Publication ethics · Research integrity · Translation

Despite an increased interest in the problem of plagiarism over the last decade in the published philosophical research literature, little attention has been given to the pernicious effects of *translation* plagiarism in published books and articles of academic philosophy. This omission should not be surprising; the identification of cases of translation plagiarism is difficult, and presenting evidence of it can be challenging. Furthermore, receptive venues for considering the problem are few. Nevertheless, open discussions of translation plagiarism can contribute to improving the integrity of scholarship in the discipline of philosophy.

Acts of translation plagiarism are largely immune to the forensic work of standard text-matching software, and they can easily be missed by even the most careful readers. Some advances in computer science have been able to detect some instances of translation plagiarism, but these advances are not yet part of standard text-matching software (Gipp 2014; Franco-Salvador et al. 2016; Sánchez-Vega et al. 2019). The problem is not restricted to humanities disciplines; one researcher reports that the undetectability of translation plagiarism by standard text-matching software results in "a large percentage of disguised scientific plagiarism going undetected" (Gipp 2014: xxiii; see Zhang 2016: 39; Weber-Wulff 2014: 92).

2.1 Defining Translation Plagiarism

Translation plagiarism—an especially subtle variety of disguised plagiarism—has been defined as "the conversion of text from one language to another with the intention of hiding its origin" (Gipp 2014: 11; see Tauginienė et al. 2018: 43; Weber-Wulff 2014: 7–8; Turell 2008: 271). This conversion of a text to another language creates a distance between original and copy that many readers may fail to overcome. The class of deceived readers for published acts of translation plagiarism initially includes the original peer reviewers and journal editors. Pre-publication peer review has been described as a particularly weak defense for detecting plagiarizing manuscripts submitted to journals (Martin 2012, 2013; De Silva and Vance 2017: 89–91). After publication, the class of deceived readers is expanded to include all those who encounter the plagiarizing article in its issued version of record and who regard it as trustworthy. Many readers—from peer reviewers to journal readers—are underprepared to discern that a plagiarizing work has already appeared in print in another language under different authorship. Some of these deceived readers, trusting the plagiarizing article, may unwittingly cite it as reliable in their own publications, further embedding the plagiarizing article in the body of published research literature.

An act of translation plagiarism is even less likely to be discovered if it copies a source from a different, yet related, discipline. For example, a successful act of translation plagiarism might involve copying from an article from a French rhetoric journal and then publishing a translated version in an English literary studies journal. In such cases, the discovery of the translation plagiarism would typically be restricted to those who have familiarity with the published literature in more than one discipline and have a facility for doing research in more than one language.

2.2 The Prevalence of Translation Plagiarism

The extent of translation plagiarism in the body of published research literature is difficult to assess. Plagiarism in all its varieties is commonly listed with falsification and fabrication to constitute the three major classes of research misconduct, and together the three are the primary reasons why journals issue retractions (Marcus and Oransky 2017; Teixeira da Silva and Dobránszki 2017). Very rarely, however, will a retraction statement identify translation plagiarism in particular as a reason for retraction. It is likely that many successful acts of translation plagiarism have gone undetected in academic journals across disciplines. Some commentators report that there are "'scholars' who simply translate documents and pass them off as their original writings" to create a research profile (Nelson Kiang, cited in Zhang 2016: 150). Again, the prevalence of such 'scholars' in the research community, and the fields that they might thrive in, is not easy to assess.

A proposed way of identifying acts of translation plagiarism is to focus on the manner that references are ordered and presented in a manuscript, rather than focusing on the words of the main text. Explicit word-for-word correlations in the main text will generally not be preserved in a translation and therefore cannot serve as a clear basis for an electronic comparison of texts (Gipp 2014). Alternative approaches to identifying translation plagiarism are being explored in the field of computer science with some success (see Franco-Salvador et al. 2013, 2016).

Acts of translation plagiarism are especially pernicious violations of research integrity. Many years—even decades—may pass between initial publication and the eventual discovery of the plagiarism. During this time, the plagiarizing article may have accrued many citations, thereby contaminating the downstream research literature. That much time often passes between publication and discovery should not be surprising, since the plagiarizing article is in a different language than its source, and a one-to-one word correspondence between original and copy is not always preserved given the syntactical differences between languages as well as the interpretive license of translators. Even when acts of translation plagiarism are discovered by readers, presenting evidence of the plagiarism in a demonstrable way may be challenging, and finding a receptive forum is not always assured. Plagiarism whistleblowers are often subject to backlash and reprisals after reporting evidence of suspected research misconduct (Fox and Beall 2014). Despite these obstacles, however, the identification of acts of translation plagiarism and the subsequent retraction of such plagiarizing articles are necessary for improving the integrity of the published research literature.

2.3 A Case of Translation Plagiarism

To exhibit the detrimental effect of translation plagiarism upon scholarly communication, and to show the difficulty in demonstrating instances of it, this chapter analyzes one case of translation plagiarism that appeared as a research article in the pages of an international journal—*Studies in Communication Sciences* (*SComS*)—in its inaugural volume. The author of record for the article (hereafter, "N.") held the office of Executive Editor at the journal at the time the article was published and was later credited by the journal as the one "who took care of running the journal from its inception" (SComS 2001: 1; Rigotti 2003: 5).

Published in the English language in 2001 and titled "Rationality as a Condition for Intercultural Understanding," the article appears in the peer-reviewed "Full Papers" section of the journal, and it purports to be an original account of the role of rationality across cultures (N. 2001). The article (hereafter, "N. 2001") consists of 17 pages of text and is followed by a two-page bibliographical reference list. The analysis of the evidence of translation plagiarism here is meant to exhibit the various ways in which translation plagiarism in academic publications corrupts scholarly communication. The evidence should be understood to be representative, rather than exhaustive, of instances where texts from N. 2001 overlap with previously published texts in a non-English language written by others.

2.4 Overlap with Peter Stemmer's *Platons Dialektik*

N. 2001 begins with a few pages that tout an increased interest in the topic of intercultural communication; the article then turns to what appears to be its main concern in Sect. 2.2, which is titled "The controversy surrounding the limited validity of rationality" (N. 2001: 84). The early part of Sect. 2.2 gives every indication of being original, theoretical work. It seems, however, to substantially overlap with texts found on the first three pages of the first chapter of a monograph in German by philosopher Peter Stemmer that was published nine years earlier (hereafter, "Stemmer 1992"). Titled *Platons Dialektik: Die frühen und mittleren Dialoge*, Stemmer 1992 is an exegesis of the use of dialectic as exhibited in the early and middle Platonic dialogues. That book begins with an analysis of Plato's social context, describing the fifth century B.C.E. as a crisis period (Krisenzeit) and offering detailed historical support, including a quotation from the Greek historian Herodotus. Table 2.1 displays an example of parallel texts between N. 2001 and Stemmer 1992, and the overlap is highlighted.

Nearly every reader will think that this section of N. 2001 is presenting an original argument about a crisis period in fifth-century Greece (with accompanying implications for the concept of rationality). The seemingly original argument depends on pieces of evidence marshalled to support the view that there was a dissolution of the traditional ethical order during this period of ancient Greek history. This evidence, however, has already appeared in print in Stemmer's book. The specific claims about international trade, travels for education and research, and encounters with other cultures—set forth in support of the thesis that these culminated in ethical relativism and skepticism in daily life—have already been published together at the beginning of the earlier 1992 book. Even though Stemmer 1992 is in the German language and N. 2001 is in English, the parallel between both texts across languages is manifest in a side-by-side comparison. The overlap in some portions is very close, with shared parallel text, yet for other sections there appears to be paraphrase or more divergence between the original and the copy.

There is no reference anywhere in N. 2001 to Stemmer 1992. No quotation marks or in-text citations in N. 2001 guide the reader to Stemmer 1992 to indicate any dependency on the earlier work. Stemmer's book is not among the 31 entries forming the two-page reference list that completes the article. In addition to the overlap of words, some idiosyncratic details support a finding of an undisclosed dependency of N. 2001 upon Stemmer 1992. For example, the quotation from Herodotus in N. 2001 and Stemmer 1992 begins and ends in the same place, and the idiosyncratic ellipses ([…]) excise the exact same portion of text in both.

With this case of translation plagiarism, a corruption of scholarly communication has occurred. Readers of N. 2001 are unknowingly encountering Stemmer 1992 through the proxy of N. 2001. Stemmer 1992 has been conscripted into plagiarizing service in N. 2001, and no credit is given to Stemmer for his words, his array and ordering of evidence, or his thesis about fifth-century Greece. A book from the discipline of philosophy by one author in one language has been incorporated into an article in the field of communication studies in another language with credit given

2.4 Overlap with Peter Stemmer's *Platons Dialektik*

Table 2.1 N. 2001 and Stemmer 1992 on the fifth-century crisis period/Krisenzeit

N. 2001: 84–85	Stemmer 1992: 4–6
A pertinent example of this can be taken from Greek antiquity during the second half of the 5th century B.C. For us, this controversy is of interest since it is accompanied by the denial of a universal claim in connection with understanding as well as the statement of a mere regional validity of rationality. When viewed from a societal perspective, the last third of the 5th century, i.e., the time when Plato was growing up, was a crisis period — a time when the dissolving of the traditional ethical order, begun in the 6th century B.C., was concluded: the traditional form of private and public life lost both its matter-of-factness as well as its unquestionable binding force. Besides the Persian wars, other possible reasons for this were the rise of international trade as well as travels for the purpose of education and research, travels which gave the Greeks an instant acquaintance with other cultures and ways of life. Experiencing that a behavior which was despised at home was accepted and even commanded in another place also caused a deep insecurity. Such insecurity resounds clearly in Herodotus's description of the funeral customs of the Greek and Indian calatier: «When Dareios was king, he once summoned all the Greeks in the vicinity to come to him, asking them to name the price for which they would be willing to consume the corpses of their fathers. They answered that no amount of money would be able to persuade them to do so. Whereupon Dareios summoned the Indian calatier, who eat the corpses of their parents, and, in the presence of the Greeks [...] asked them to name the price for which they would cremate their dead fathers. They shrieked and implored him earnestly to desist from such godless utterances»⁵ [Herodot: *Historiae*: in 4 vol. With an English translation by A. D. Godley. London: Heinemann (The Loeb Classical Library), Vol. I, Book III, 38, 3-4.] The consequences which could be drawn from such insights as to the limited validity of allegedly generally valid values were of a different nature still. One could deduce from this that one's own tradition represents only one of many possible conventions, which would mean it could be ascribed only a conditional value. Other Greeks tried to measure the different cultural traditions against one common criterion; for example, one such criterion taken into consideration above all was reasonableness. By means of reason — understood as the source of knowledge of cosmic, divine things — the suitability of one's own tradition could be tested; in part also its superiority over other life forms could be proved. However, from the beginning, the attempt to cope theoretically with this crisis by a return to the claim of the universality of reason was countered by the demands of those who accepted the differences in customs as decisive evidence for a legitimate ethical relativism and scepticism practiced in daily life.	Die zweite Hälfte des 5. Jahrhunderts war eine Krisenzeit, in der sich die schon im 6. Jahrhundert einsetzende Auflösung der überkommenen ethischen Ordnung beschleunigte und in der Realität dieses Zerfalls auch zunehmend wahrgenommen und als bedrohlich empfunden wurde. Die traditionelle Gestalt des privaten und öffentlichen Lebens verlor ihre Selbstverständlichkeit und ihre fraglose Verbindlichkeit. Besonders im letzten Drittel des Jahrhunderts, in der Zeit also, in der Platon aufwuchs und seine intellektuelle Formung empfing, wurde deutlich, daß die tradierten ethischen, politischen und religiösen Orientierungen ihre selbstverständliche Geltung bereits verloren hatten oder dabei waren, sie zu verlieren. (...). Durch die Perserkriege, den aufkommenden internationalen Handel, durch Bildungs- und Forschungsreisen erhielten die Griechen in dieser Zeit Kunde von der Verschiedenheit der Sitten und Bräuche bei anderen Völkern. Ihre eigene Welt war nicht mehr die Welt schlechthin. Und die Erfahrung, daß anderswo ein Verhalten akzeptiert, ja sogar geboten war, das zu Hause völlig außerhalb des Denkbaren lag, war erschütternd. Man kann diese tiefe Verunsicherung spüren, wenn Herodot über die verschiedenen Bestattungsbräuche bei Griechen und indischen Kalatiern berichtet: „Als Dareios König war, ließ er einmal alle Griechen seiner Umgebung zu sich rufen und fragte sie, um welchen Lohn sie bereit wären, die Leichen ihrer Väter zu verspeisen. Die aber antworteten, sie würden das um keinen Preis tun. Darauf rief Dareios die indischen Kalatier, die die Leichen der Eltern essen, und fragte sie in Anwesenheit der Griechen ..., um welchen Preis sie ihre verstorbenen Väter verbrennen möchten. Sie schrieen laut auf und baten ihn inständig, solch gottlose Worte zu lassen."² [Herodot III, 38,3–4 (Übers,: J. Feix); vgl. Auch I, 216, 1–4.] (...). Einsicht in die nur regionale Geltung von vermeintlich allgemein gültigen ethischen Wertvorstellungen (...). Aus dieser neuen Sicht ließen sich verschiedene Konsequenzen ziehen: Man konnte den menschlichen Nomos als bloß beliebige, willkürliche Konvention insgesamt verwerfen und sich nur an das halten, was man für nicht-konventionell, für nicht-variabel, nämlich natürlich hielt. Man konnte ebenso versuchen, die verschiedenen kulturellen Traditionen an einem Kriterium zu messen, das es erlaubte, eine Tradition den anderen vorzuziehen. Dieses Kriterium fand man entweder in der Naturgemäßheit, in der Vernünftigkeit ouer der Glückszuträglichkeit. (...) Doch diesseits aller Versuche, die Krise theoretisch zu meistern, bedeutete die Erfahrung der Verschiedenheit der Sitten zunächst eine Labilisierung der ererbten Moral und infolgedessen einen wachsenden, im tagtäglichen Tun realisierten ethischen Relativismus und Skeptizismus.

to someone else. The repurposing of text is concealed in at least three ways: first, by the transfer of the text from one distinct discipline to another; second, by a rendering of the original German text into English; and third, by the absence of any reference to Stemmer 1992, the source text.

2.4.1 The Role of Disguising References

As noted above, the varieties of disguised plagiarism—including translation plagiarism—are so designated because they each exhibit some concealing activity that creates distance between original and copy. The undisclosed dependency of N. 2001 on Stemmer 1992 is further obscured in subtle ways beyond the change in language. For example, a detailed footnote in N. 2001 claims that the text from Herodotus that is being cited is from the 1938 English translation by A. D. Godley from the Loeb Classical Library series. Despite this attribution, the text is not in fact from the Godley translation. A closer inspection shows that the word order of the text from Herodotus in English as it is given in N. 2001 matches the German version cited by Stemmer, who himself credits Josef Feix for the German translation of Herodotus in a footnote. N. 2001's reference to Godley appears to be an act of obfuscation, directing the reader away from its real source. Indeed, in the N. 2001 footnote, the reference to Godley is supplemented with the information "Vol. I, Book III, 38, 3–4"; but that volume number and those section numbers match the Feix translation and not the Godley translation. (Book III, Chap. 38 in Herodotus occurs in volume 2—not 1—in the Godley translation, and there are no section numbers in Godley's translation.) The volume and section numbers given in N. 2001 inexplicably attributed to Godley do, however, correspond exactly to what is found in Stemmer in his footnote that credits the translation to Feix. These details concerning the misdirecting footnote further strengthen the claim that N. 2001 is surreptitiously presenting material from Stemmer 2001.

There is another oddity: the reference list of N. 2001 credits a different English translation for Herodotus: a 1962 English translation by Harry Carter. But the English text of Herodotus presented in N. 2001 is neither from Godley nor Carter, despite these explicit and implied attributions. An example offered in Table 2.2 shows that the text of Herodotus as given in English translation in N. 2001 cannot be that of Godley (Herodotus 1938) or Carter (Herodotus 1962) but appears to be an English rendering of the German translation by Feix (Herodotus 1980) that is cited in Stemmer 1992.[1]

Table 2.2 A selection of Herodotus translations

Stemmer 1992: 5 (Citing Feix, 1980: 216)	Darauf rief Dareios die indischen Kalatier, die die Leichen der Eltern essen, und fragte sie in Anwesenheit der Griechen . . ., um welchen Preis sie ihre verstorbenen Väter verbrennen möchten.	✓
N. 2001: 85	Whereupon Dareios summoned the Indian calatier, who eat the corpses of their parents, and, in the presence of the Greeks [. . .] asked them to name the price for which they would cremate their dead fathers.	✓
Godley 1938: 51	Then he summoned those Indians who are called Callatiae, who eat their parents, and asked them (the Greeks being present and understanding by interpretation what was said) what would make them willing to burn their fathers at death.	✗
Carter 1962: 182	Then he sent for some Indians, called Callatians, who eat their parents, and in the presence of the Greeks, who were told by interpreters what was being said, he asked the Indians how much money would persuade them to burn their fathers after death	✗

---

[1] I am grateful to Pernille Harsting for discovering that the English word order of the text of Herodotus in N. 2001 matches the German translation of Herodotus by Feix.

2.4 Overlap with Peter Stemmer's *Platons Dialektik*

The general word order, sentence structure, and use of ellipses are common to N. 2001 and Stemmer 1992, but not to the published English translations by Godley or Carter.

Still further, N. 2001 presents proper names with standard German orthography (e.g., "Herodot," "Dareios")—as they appear in Stemmer's book and in Feix's translation—rather than with standard English orthography (e.g., "Herodotus," "Darius") that is found in English translations including Godley and Carter. In short, the translations by Godley and Carter are not relevant to N. 2001 despite a footnote and a bibliographical entry to the contrary.

2.4.2 The Hidden German Phase of N. 2001

Nearly every reader encountering N. 2001 will take the text about a crisis period in fifth-century Greece to be an original analysis. Based on the text alone, as it appears in N. 2001, readers will not be able to reduce it to its hidden source in Stemmer 1992. Figure 2.1 illustrates how a typical reader will encounter the argument in N. 2001 involving Herodotus.

In short: to the typical reader it will appear that there are only two phases: a Greek phase and an English phase. To such a reader it will appear that the Greek text of Herodotus (1) translated by Godley (2) has been incorporated into an original argument in N. 2001 (3). The typical reader will trust that the footnote offering credit to Godley is a genuine act of attribution to a source. Furthermore, in the absence of any quotation marks or citations to Stemmer 1992, the reader will believe that the argument and analysis surrounding the quotation of Herodotus are original to N. 2001 and will not infer that a significant portion of N. 2001 originates in a previously published and unreferenced scholarly book. Furthermore, the reader will assume that the argument and analysis involve only two languages, as the Greek text of Herodotus passes into English via Godley and then is incorporated into the English argument as present in N. 2001.

The evidence offered here shows, however, that the reader drawing such a conclusion is misled. The real situation is much more complex, as there is a disguised

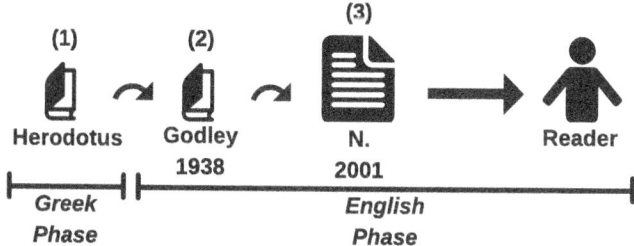

Fig. 2.1 How typical readers encounter N. 2001's account of Herodotus

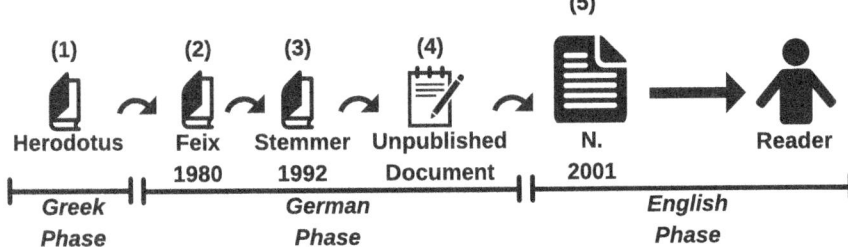

Fig. 2.2 Disguised textual transmission in N. 2001

textual transmission involving a third language—German—and *this German phase has been entirely obscured to the reader*. Figure 2.2 illustrates the reconstruction of the disguised textual transmission, displaying the concealed German phase in the transmission of the text.

The actual path from Herodotus to reader will have begun with the translation of the Greek text of Herodotus (1) into a German translation by Feix (2), a portion of which is then extracted in Stemmer's book (3) with appropriate credit by means of quotation marks and a footnote citation. Stemmer's extract of Herodotus and his surrounding argument and analysis is then fashioned into a plagiarizing unpublished German document (4), which is then translated into English and published as N. 2001 (5) without reference to Stemmer 1992. When the reader encounters N. 2001 in print, therefore, the entire German phase (2)-(3)-(4) of the construction of N. 2001 has beenconcealed: the reader knows nothing of the translating work by Feix (2), the text as originally written by Stemmer (3), or the plagiarizing German unpublished document (4).

2.4.3 The Hypothesized Document of the Hidden German Phase

The elements of the process depicted in Fig. 2.2 are all known to exist except for one. That is, N. 2001 (5), Stemmer 1992 (3), Feix 1980 (2), and a Greek text of Herodotus (1) can all be confirmed to exist. Only one part is hypothesized: that there exists (or existed) a German unpublished document (4) at the end of the German phase. This hypothesized German unpublished document would be one in which the following conditions are met:

(a) it contains a German text that generally corresponds with the selection of N. 2001 about the fifth-century crisis period (Krisenzeit) given above in Table 2.1;
(b) the texts of Stemmer show up in it without attribution in German (rather than in English translation) and the overlap is verbatim and near verbatim in many instances;

2.4 Overlap with Peter Stemmer's *Platons Dialektik*

(c) the unpublished document presents Feix's German translation of Herodotus in the same way it is presented by Stemmer; and
(d) the superfluous footnote to Godley's English translation of Herodotus will not be present, but instead Stemmer's footnote to Feix's German translation of Herodotus will be reproduced.

It is difficult to confirm the existence of a hypothesized German unpublished document that meets this these particular conditions.

Nevertheless, a portion of such a hypothesized German unpublished document seems to have been published by N. two years later in German in 2003, and its public existence confirms the analysis defended here (N. 2003). A German-language article titled "Möglichkeiten und Grenzen einer 'Ethik der Normalität des Fremden'" purports to offer an analysis of a book by Hans Hunfield, a philosopher at Katholischen Universität Eichstätt-Ingolstadt. What is important for the argument here is that the antepenultimate and penultimate pages of that article contain a section that appears to be the German text for the section of N. 2001 that was exhibited in Table 2.1. This section from N. 2003 is shown in Table 2.3, which shows the identity of N. 2001 and

Table 2.3 N. 2003 as the German text for N. 2001, with Stemmer 1992 in bold

N. 2001: 84–85	N. 2003: 61–62
the last third of the 5th century, i.e., **the time when Plato was growing up, was a crisis period** — a time when the dissolving of the traditional ethical order, begun in the 6th century B.C., was concluded; **the traditional form of private and public life lost both its matter-of-factness as well as its unquestionable binding force. Besides the Persian wars,** other possible reasons for this were the rise of international trade as well as travels for the purpose of education and research, travels which gave the Greeks an instant acquaintance with other cultures and ways of life. Experiencing that a behavior which was despised at home was accepted and even commanded in another place also caused a deep insecurity. Such insecurity resounds clearly in Herodotus's description of the funeral customs of the Greek and Indian calatier: «When Dareios was king, he once summoned all the Greeks in the vicinity to come to him, asking them to name the price for which they would be willing to consume the corpses of their fathers. They answered that no amount of money would be able to persuade them to do so. Whereupon Dareios summoned the Indian calatier, who eat the corpses of their parents, and, in the presence of the Greeks [...] asked them to name the price for which they would cremate their dead fathers. They shrieked and implored him earnestly to desist from such godless utterances»[5] [Herodot: *Historiae*: in 4 vol. With an English translation by A. D. Godley. London: Heinemann (The Loeb Classical Library), Vol. I, Book III, 38, 3-4.] The consequences which could be drawn from such insights as to the limited validity of allegedly generally valid values were of a different nature still. One could deduce from this that one's own tradition represents only one of many possible conventions, which would mean it could be ascribed only a conditional value. Other Greeks tried to measure the different cultural traditions against one common criterion; for example, one such criterion taken into consideration above all was reasonableness.	Das letzte Drittel des 5. Jahrhunderts, **die Zeit also, in der Platon aufwuchs, war** gesellschaftlich gesehen **eine Krisenzeit, in der sich die im 6. Jahrhundert** beginnende Auflösung der überkommenen ethischen Ordnung vollends vollzog: **Die traditionelle Gestalt des privaten und öffentlichen Lebens verlor** ebenso ihre Selbstverständlichkeit wie **ihre fraglose Verbindlichkeit.** Als mögliche Ursachen dafür sind neben den **Perserkriegen** das Aufkommen des internationalen Handels sowie Bildungs- und Forschungsreisen zu nennen, Reisen, durch die die Griechen rasch andere Kulturen und Lebensformen kennenlernten. In die damit einhergehende Erfahrung, dass anderswo ein Verhalten akzeptiert, ja geboten war, das zu Hause als verpönt galt, mischte sich eine tiefe Verunsicherung. Sie klingt noch deutlich nach in Herodots Schilderung der Bestattungsgebräuche bei Griechen und indischen Kalatiern: "Als Dareios König war, ließ er einmal alle Griechen seiner Umgebung zu sich rufen und fragte sie, um welchen Lohn sie bereit wären, die Leichen ihrer Väter zu verspeisen. Die aber antworteten, sie würden das um keinen Preis tun. Darauf rief Dareios die indischen Kalatier, die die Leichen der Eltern essen, und fragte sie in Anwesenheit der Griechen [...], um welchen Preis sie ihre verstorbenen Väter verbrennen möchten. Sie schrieen laut auf und baten ihn inständig, solch gottlose Worte zu lassen." Die Folgen, die man aus der **Einsicht in die** begrenzte Geltung von vermeintlich allgemein gültigen Wertvorstellungen ziehen kann, sind noch unterschiedlicher Art. Man kann daraus etwa ableiten, dass die eigene **Tradition** nur eine beliebige **Konvention** unter anderen möglichen darstellt, weshalb ihr auch nur ein bedingter Wert zukommt. Ein anderer Weg besteht darin, **die verschiedenen kulturellen Traditionen an einem** übergreifenden **Kriterium zu messen**. Als ein solches kommt neben anderen vor allem das **Kriterium der Vernünftigkeit**, der Rationalität in Betracht. [8 **Herodot**: *Historien*. Griechisch-Deutsch. Hrsg. v. Josef **Feix**. 3. durchgeseh. Auflage (München u.a. 1980), Bd. I, **III. Buch, 38, 3-4.**]

N. 2003 through highlighting, with overlap with Stemmer 1992 given in bold. On the whole, this passage satisfies the conditions identified as a–d above.

The German text of N. 2003 largely overlaps with its English counterpart in N. 2001. It still gives no reference to Stemmer 1992 but contains verbatim and near verbatim texts from Stemmer 1992. There is no superfluous footnote to Godley's translation of Herodotus, but instead there is an accurate reference to Feix's translation (just as there is in Stemmer 1992). Again, the idiosyncratic ellipses are present in the quotation from Herodotus in N. 2003 (just as they are present in Stemmer 1992 and in N. 2001). This N. 2003 text confirms the disguised textual transmission that was illustrated in Fig. 2.2.

In addition to the direct evidence of translation plagiarism shown in the parallel texts exhibited in Table 2.1, and the confirmation of a hypothesized plagiarizing German unpublished document in Table 2.3, there is further independent circumstantial evidence that the author of record for N. 2001 was familiar with Stemmer's monograph prior to the publication of N. 2001. In short: *the author of record for N. 2001 demonstrates prior familiarity with Stemmer 1992 by discussing it explicitly in a book published earlier, in 2000* (N. 2000: 33, 332). This prior discussion would seem to rule out an exculpatory argument that would posit a simultaneous discovery of what is substantially the same text by two different people writing in different languages nine years apart. One cannot reasonably maintain that the overlap of text is simply due to wholly independent and unrelated fortuitous discoveries of the same argument by two people: first by Stemmer in 1992 and then by N. in 2001 (and again later in N. 2003). The public evidence that the author of record for N. 2001 was familiar with Stemmer 1992 prior to the publication of N. 2001 rules out such a view.

2.4.4 Relay Translation Plagiarism

This instance of translation plagiarism in N. 2001 appears to deny credit not only to Stemmer but also to Feix, since Feix's German translation of Herodotus is apparently the basis for the English translation in N. 2001, despite the misdirecting footnote reference to an English translation by Godley. The translation of a work from a translation rather than from its original is sometimes designated as "relay translation" or "indirect retranslation."

Some theorists have argued that acts of relay translation or indirect retranslation can constitute acts of plagiarism when credit is not given to the first translation that serves as the basis for the second translation. In the words of one theorist: "If the retranslator adopts some of the lexical, syntactic or stylistic units from previous translations, this effort is evidently a case of plagiarism" (Şahin et al. 2015: 198; see also Şahin et al. 2019; Jianzhong 2003; Turell 2008). Translation plagiarism of this kind has been chronicled for philosophical works including Immanuel Kant's *Critique of Pure Reason* and Thomas More's *Utopia* (Kanra 2019; Elgül 2019). In

2.4 Overlap with Peter Stemmer's *Platons Dialektik*

the case here, Feix's translation of Herodotus appears to have been used in an act of relay translation, but without credit.

2.4.5 Duplication in the Downstream Literature

Plagiarizing articles not only deny credit to original authors, but they also create duplications in the published research literature, since a plagiarizing article creates a second yet disguised path by which readers can reach the content produced by the original author. One group of readers encounters the content directly and efficiently by reading it as it is first published under the name of the original author. Another group of readers, however, encounters the content indirectly and inefficiently through the proxy of a plagiarizing article that conceals the true origin of its content. Figure 2.3 illustrates these two paths for the case of N. 2001's overlap with Stemmer 1992.

Along the first path, readers encounter the content of Stemmer's book through the book itself, as it is published under Stemmer's name. Such readers draw no false conclusions about the origin of book's content. Along the second path, however, readers encounter the content of Stemmer's book in a hidden or disguised way in N. 2001, not knowing that it re-presents texts from Stemmer without attribution. Since N. 2001 does not reference Stemmer 1992 in any way, and since the footnote to Godley's 1938 translation points readers in the wrong direction, those on the second path are disadvantaged in trying to enter into a scholarly interchange of ideas. This disadvantage is most acute when some of the readers on the second path respond to N. 2001 in print in their own publications, not knowing that they are engaging Stemmer's book through the proxy of N. 2001.

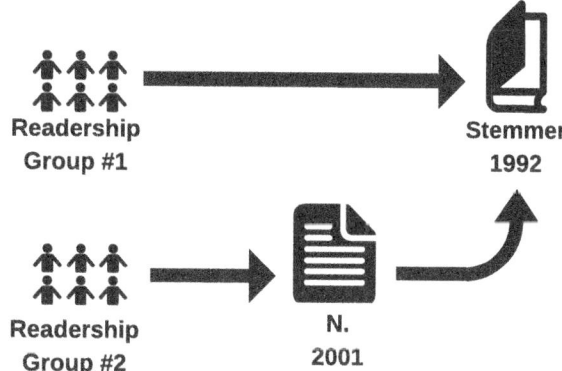

Fig. 2.3 N. 2001 as a proxy for Stemmer 1992

2.4.6 Corruptions of the Downstream Literature

If what is being argued here is correct, then many of the citations to N. 2001 in the subsequent published literature should really be to Stemmer 1992. Authors who respond to N. 2001 in print are in the dark about who their real interlocutor is. Through no fault of their own, the authors of articles that engage N. 2001 fundamentally misidentify the author they are addressing, creating a fundamental corruption of scholarly communication. This point is often overlooked in discussions about the harm of plagiarism.

A plagiarizing article thereby functions as a Doppelgänger that draws researchers away from engaging genuine authors directly. One must assume that all the authors who have subsequently cited N. 2001 in print in their own publications have been unaware of the overlap between N. 2001 and Stemmer 1992 and have cited the later article in good faith. An example can show how translation plagiarism damages the quality of scholarly communication in the downstream literature. In the 2009 book *Global Linguistics*, authors Marcelo Danesi and Andrea Rocci discuss N. 2001 in positive terms, summarizing its discussion of Herodotus, and even providing a quotation from N. 2001:

> When the ancient Greeks began to come into contact with the Persians and other Asian peoples, they experienced a deep cultural shock as they discovered that 'a behavior which was despised at home was accepted and even commanded in another place' ([N.] 2001: 85). Herodotus (5th century BC), a Greek historian that can be considered also the first ethnographer, widely comments on the shocking diverse habits of different peoples, offering a testimony of this discovery. As [N.] (2001) observes, the quest for reasonableness and universality that characterizes ancient Greek philosophy initially emerged also as an attempt to cope with this cultural shock (Danesi and Rocci 2009: 11).

The quotation that Danesi and Rocci attribute to N. 2001 is really found in Stemmer 1992 (see Table 2.1). Furthermore, the interpretation of Herodotus and the account of Greek rationality they attribute to N. 2001 is also in Stemmer 1992. One must draw the conclusion that Danesi and Rocci are not aware that they are unwittingly engaging Stemmer in print through the proxy of N. 2001. One should not, of course, blame authors who unknowingly engage and cite plagiarizing articles. The default assumption of scholars is that every author of record for scholarly works is the genuine author of what appears and print and that authors who publish scholarly works will use the standard conventions (e.g., quotation marks and accurate citations) for indicating what is original and what is not in scholarly writing. In this case of translation plagiarism, not only is Stemmer not credited for his contribution, but the quality of the subsequent responding literature is lessened as Danesi and Rocci unwittingly engage someone other than Stemmer, who is the hidden genuine author of a portion of N. 2001. As responding articles continue to credit N. 2001 for the content of Stemmer 1992, the downstream literature becomes less reliable and the quality of scholarly communication is vitiated.

The positive discussion of N. 2001 by Danesi and Rocci is not an outlier. In a 2009 monograph, Eddo Rigotti also cites the article positively, engaging N. 2001 on the interpretation of Herodotus, again without any indication of any awareness that

2.4 Overlap with Peter Stemmer's *Platons Dialektik*

N. 2001 is presenting texts of Stemmer in English translation without attribution. Rigotti engages the discussion of Herodotus as it appears in N. 2001, stating:

> Erodoto sembra consapevole della composizione multiculturale dell'impero di Ciro, ponendo embrionalmente la tematica della comunicazione interculturale. Ma vedi in proposito: [N.], 'Rationality as a Condition for Intercultural Understanding,' in «Studies in Communication Sciences» (Rigotti 2009: 51).

This interchange in the downstream literature where Rigotti engages N. 2001 is especially remarkable when one considers that Rigotti himself was the Editor-in-Chief of *SComS* at the time N. 2001 was published. Rigotti served alongside N. who was (as noted above) the Executive Editor for the inaugural volume of the journal (SComS 2001: 1).

2.5 Overlap with Stefan Gosepath's *Aufgeklärtes Eigeninteresse*

The preceding section has offered evidence of translation plagiarism, arguing that there is textual overlap between the beginning of the second section of N. 2001 and a German monograph by Peter Stemmer. In total, however, the textual overlap with Stemmer 1992 occurs on only two of the 19 pages that constitute N. 2001. To account for the substance of the article, one must look elsewhere. The main source for the article—covering major portions of sections 2, 3, and 4 of N. 2001—is a different German philosophical monograph. Authored by philosopher Stefan Gosepath and published in 1992, the large 420-page volume *Aufgeklärtes Eigeninteresse: Eine Theorie theoretischer und praktischer Rationalität* [Enlightened Self-Interest: A Theory of Theoretical and Practical Rationality] presents a detailed account of the concept of rationality (hereafter "Gosepath 1992"). Much of N. 2001 is an English rendering of key parts of that book.

Unlike Stemmer 1992, which is nowhere referenced or cited in N. 2001, Gosepath 1992 is one of the 31 entries constituting the article's two-page reference list. Yet there are no quotation marks, extract quoting, in-text citations, or any other conventions in use in the body of the article that would indicate to the reader that N. 2001 is repackaging, in English, key portions of Gosepath's German philosophical monograph. Large swaths of text from Gosepath 1992 show up unacknowledged—in English translation—on at least 12 of the 19 pages of the article. If the texts in N. 2001 that overlap with Gosepath 1992 are removed, little content of significance remains to sections 2–4 of the article.

The texts in N. 2001 overlap with portions from a wide array of sections in Gosepath 1992. Table 2.4 presents an excerpt from N. 2001 with a parallel text found in the first chapter of part one of Gosepath 1992 (Teil I: Allgemeine Begriffsbestimmung/Kapitel I. Eine begriffliche Landkarte). The passage is a key one in Gosepath's project, for it presents Gosepath's account of what it means for an opinion to be rational.

Table 2.4 N. 2001 and Gosepath 1992 on whether an opinion/Meinung is rational/rational

N. 2001: 90–91	Gosepath 1992: 27
an opinion can be called rational only when the opinion's proponent has justifiable reasons for it. Thereby, the ability to justify an opinion is not solely a matter of deductive or inductive reasoning, but of a general ability to justify statements, i.e., to give an account of them. Both the Greek expression «λόγον διδόναι» as well as the Latin «rationem reddere» suggest this meaning, too.[11][... Cf. the articles «Denken» and «Rechnen» in *Hist. Wörterbuch der Philosophe* 2 (1972): 64-66] And the corresponding translation «reason», «raison», or «ragione» means, respectively, «a reason» and «reason», An opinion is therefore substantiated by reason when it can be derived logically correctly from the reasons justifying it. In general: «*S*'s opinion at the time *t* is rational and reasonable when *S* at the time *t* has good reason(s) to be of the opinion that *P*»	Eine Meinung ist rational, wenn derjenige, der sie hat, über rechtfertigende Gründe verfügt. Die Fähigkeit, eine Meinung zu rechtfertigen, ist nicht nur ein Vermögen des deduktiven Schließens, sondern allgemeiner die Fähigkeit, Aussagen zu begründen, auszuweisen, d. h. für sie Rede und Antwort stehen zu können (lateinisch: rationem reddere, griechisch: logon didonai). »Ratio«, »raison«, »reason« bedeuten ebensosehr »Grund« wie »Vernunft«. Die Minimalbedingung für eine akzeptable Begründung ist die logisch korrekte Ableitung einer Meinung aus den sie rechtfertigenden Gründen [...]. Allgemeiner gilt daher: Eine Meinung von S zu t ist rational oder vernünftig, wenn S zu t gute Gründe hat zu meinen, daß p.

Gosepath finds evidence for his account of rationality in the etymology of Greek and Latin cognates, and this philological evidence inexplicably appears also in N. 2001 without attribution. Furthermore, Gosepath's distinctive terminology—including variable *S* for the opiner and variable *P* for the opinion—and his definition that "Eine Meinung von S zu t ist rational oder vernünftig, wenn S zu t gute Gründe hat zu meinen, daß p" appears almost word-for-word in English translation in N. 2001.

There is one significant addition in the selection of N. 2001 shown in Table 2.4 that has no corresponding text in Gosepath 1992, however. A footnote reference directs the reader's attention to *Historisches Wörterbuch der Philosophie* (*HWPh*) for the etymological foundation of the analysis. This footnote reference in N. 2001, nevertheless, seems gratuitous since the consideration of Greek and Latin cognates seems to derive immediately from Gosepath 1992 rather than from any reference work such as *HWPh*. The reader's attention is misdirected when credit is given to *HWPh* for a text that copies Gosepath 1992. Much like the citation to the Godley translation mentioned above, the reference to *HWPh* points the reader to a text other than its source.

2.5.1 Meta-narrative in N. 2001

By omitting quotation marks, block quoting, extracting, in-text citations, and other academic conventions of attribution, a plagiarizing article generates for the reader a façade of genuine authorship. A plagiarizing article omits the use of conventional signals for indicating to readers what is original and what is not. In addition to the subtle omission of established conventions of attribution in academic writing, there are other ways in which plagiarizing articles can support an illusion of genuine authorship. One such way is the use of meta-narrative in which the plagiarizing author of record purports to explain what is being accomplished in the article. By using the first-person pronoun "I," the reader is disposed to think that the plagiarizing author of record is the genuine author of the text that follows.

2.5 Overlap with Stefan Gosepath's *Aufgeklärtes Eigeninteresse*

N. 2001 at times features meta-narrative, beginning with the abstract that begins the paper. There the reader finds such assertions as, "In this paper, I first discuss how [...]" and "I discuss [...] and argue that [...]" (81). Such meta-narrative strengthens the appearance that the content that follows is original to the author of record of the article, whose name appears on the first page right above the title. An additional instance of meta-narrative can be found later in the paper. The reader finds what appears to be the author of record explaining the use of a key term by providing a stipulative definition, stating "By cognitive condition, I mean the whole set of accepted opinions, convictions, evidence, and goals possessed by a certain person at a certain period in time" (N. 2001: 94). The use of "I mean" is strong meta-narrative, but it is simply repeating Gosepath's first person claim "Unter einem kognitiven Ausgangszustand fasse ich den Korpus der akzeptierten Meinungen, die zur Verfügung stehenden Evidenzen und die Ziele, die eine bestimmte Person zu einem bestimmten Zeitpunkt hat" (Gosepath 1992: 185). Table 2.5 presents the larger context in which the meta-narrative claim is found in both Gosepath 1992 and N. 2001. Not only is the stipulative definition given in the form of meta-narrative in both texts, but both texts employ the same technical nomenclature for expressing the specific conditions for possessing a rational opinion. Furthermore, both texts express very particularized claims, such as features of the rain dance practices of the African Azande tribe involving gods, spirits, and magicians.

There appear to be some paraphrasing, modifications in the order of sentences, and additions in the selection of N. 2001 exhibited in Table 2.5. Some of the alterations are somewhat minor but still striking; the parenthetical aside in Gosepath 1992 "(nach

Table 2.5 N. 2001 and Gosepath 1992 on the conditions of rationality

N. 2001: 94–95	Gosepath 1992: 184–185
Whether or not it is rational to have opinion *P* depends on two conditions: on the one hand, the respective cognitive condition; on the other hand, the rules of rationality. By cognitive condition, I mean the whole set of accepted opinions, convictions, evidence, and goals possessed by a certain person at a certain period in time. This cognitive state must first of all be differentiated from the rules of rationality. Both go along with reasoning even if one can separate them into individual problems in each case. This distinction may be clarified on the basis of the aforementioned study of the ethnologist Evans-Pritchard. Among other things, he describes the rain dance practices of the African native Azande tribe. These practices can be reconstructed by including the underlying cognitive state. For they incorporate the view that gods, spirits and magicians rule over the powers of nature and can cause rain. For it to rain, the gods and spirits must be aroused by befitting practices (Evans-Pritchard 1937). If one questions how the Azande can maintain these opinions in face of evidence to the contrary (as viewed by a West European), then the rules of rationality become the issue. The example shows that one is able to speak of two possible types of the relativity of rationality. Rationality is relative first of all in relation to the respective cognitive state. Secondly, there is at the same time a relativity in relation to certain rules or standards of theoretical rationality. If one discusses the problem of rationality and its relativity, then one has to be aware of the kind of relativity being referred to.	Ob es rational ist, p zu glauben, ist abhängig von einem kognitiven Ausgangszustand, von dem aus entsprechend den Regeln der Rationalität die Meinung, daß p, gebildet wurde. Entscheidend für eine Begründung sind also einmal der kognitive Ausgangszustand und zum anderen die Regeln. Dadurch ergeben sich zwei Arten von Relativität der theoretischen Rationalität: zum einen eine Relativität mit Bezug auf kognitive Ausgangszustände, zum anderen eine Relativität mit Bezug auf die sog. Regeln bzw. Standards theoretischer Rationalität. [...]. So versteht man die Praktiken des Regentanzes des afrikanischen Ureinwohner-Stammes der Zande durchaus,[7] [...Evans-Pritchard] wenn man ihnen die gleichen Rationalitätsregeln unterstellt und dann erfährt, daß sie einen anderen Ausgangszustand haben, und zwar insbesondere die Meinung, daß es Götter, Geister, Magier gibt, die den Regen beeinflussen können und sich durch rituelle Praktiken dazu bewegen lassen. Wenn man hingegen fragt, wie die Zande diese Meinungen angesichts der gegenteiligen Evidenzen (nach unserer Sicht) aufrechterhalten können, dann werden die Regeln, hier der induktiven Bestätigung, thematisiert. [...]. Unter einem kognitiven Ausgangszustand fasse ich den Korpus der akzeptierten Meinungen, die zur Verfügung stehenden Evidenzen und die Ziele, die eine bestimmte Person zu einem bestimmten Zeitpunkt hat.

unserer Sicht)" appears in N. 2001 in the expanded form of "(as viewed by a West European)" (Gosepath 1992: 195; N. 2001: 94).

2.5.2 The Repetition of Idiosyncratic Examples

Additional evidence in support of the view that Gosepath 1992 is copied by N. 2001 is the re-appearance of very specific unusual examples in the later article. Table 2.6 presents an extract of N. 2001 in which several examples of irrational opinions are given, and it is useful to compare them with their predecessors in Gosepath 1992. N. 2001 describes a situation involving an agent who searches for someone's internet address while believing the other does not possess one. If one looks to the parallel text in Gosepath, the example is similar, involving an agent who searches for a phone number while believing that the other does not possess one. Two additional examples are shared. Both N. 2001 and Stemmer 1992 discuss an agent who believes that winning the lottery is imminent, but in N. 2001 the example has been updated to indicate the lottery is paid out in Swiss francs. Both also discuss examples of voters who believe a candidate's claim of impending victory, but in N. 2001 the example has been shortened by omitting the name of a particular candidate.

Such examples shared by N. 2001 and Gosepath 1992 are particular; the overlap between them cannot be dismissed as coincidental, especially in light of the many parallels in the text surrounding the examples. Table 2.6 also exhibits that both works define the holding of an opinion with the same nomenclature and with the same variables for formalizing what it means to hold an opinion. Holding an opinion

Table 2.6 Examples common to N. 2001 and Gosepath 1992

N. 2001: 88–90	Gosepath 1992: 23–24
deliberations, opinions, or arguments may be candidates for rationality within the cognitive area, actions within the behavioral area, and desires within the emotional area [...]. When we call cognitive deliberations, discourses, arguments, or thoughts rational we are referring to them neither as linguistic constructs nor going into the respective content of their declarative statements (technically speaking, the proposition or the propositional subject matter). Such a reference is either true or false, but not rational or irrational. By using the expression «rational,» we are describing opinions in a much broader sense.In the following remarks, «opinion» should be understood as a proposition held to be true. Holding a proposition to be true includes: believing, expecting, supposing something, being convinced of something, considering its possibility, etc. So, holding an opinion is a relation between a subject (S) and a proposition (P), which can be formalized as «S is of the opinion that P...» To be exact, one has to add:S is of the opinion P at a certain point in time or during a certain period of time (t), i.e., «S is of the opinion P at time t.» [...] A series of case examples will serve as a means of examining this question more closely; in these examples, the homogenous meaning of an «irrational» opinion will be questioned. a) Searching for someone'sInternet address, although one knows for sure that the person does not have one. b) Being sure that one will win a million Swiss francs in the next lottery [...]. e) Believing that a politician will win the election on the basis of his/her own statement to that effect.	Im theoretischen Bereich, verstanden als die kognitive Sphäre,scheinen sich auf den ersten Blick eine Menge Kandidaten für Rationalität anzubieten. Beispiele dafür wären: Überlegungen, Überzeugungen, Gedanken, Reden, Argumente usw. Was ist es, was wir in diesem Bereich als rational bezeichnen? [...] Aber auch der Inhalt eines Aussagesatzes, die Proposition, kann weder rational noch irrational sein, sie ist wahr oder falsch.Als rational oder irrational bezeichnen wir umgangssprachlich vielmehr Meinungen. Unter »Meinung« soll hier das Fürwahrhalten einer Proposition verstanden werden. Wir benutzen dafür Worte wie: glauben, erwarten, annehmen, überzeugt sein, für möglich halten, für wahrscheinlich halten, als unmöglich ansehen, ahnen, auf etwas gefaßt sein, sich etwas einbilden, vermuten, mit etwas rechnen etc. [...]. Das Meinen ist eine Relation zwischen einer Person und einer Proposition zu einem Zeitpunkt oder in einem Zeitraum: S meint zu t, daß p [...]. Einige Beispiele für Meinungen, die odem Vorwurf der Irrationalität ausgesetzt sein können: (a) in einem Telefonbuch die Nummer von jemandem zu suchen, von dem man eigentlich weiß, daß er keinen Telefonanschluß besitzt [...]; (c) zu behaupten, am nächsten Samstag werde man sicher im Lotto gewinnen [...]; (f) zu glauben, die SPD werde die nächsten Wahlen gewinnen, nur weil Willy Brandt es gesagt hat.

2.5 Overlap with Stefan Gosepath's *Aufgeklärtes Eigeninteresse*

is described in both as a relation between a person and a position at a certain point in time or during a certain period of time where S is of the opinion P at t. Again, the reader has no way of knowing—at least from the text itself—that the content of N. 2001 can be found in a book published 9 years earlier in a different language.

2.5.3 Bibliographical Overlap

Another way of approaching the derivation of N. 2001 from Gosepath 1992 is to examine the parallel texts in light of shared bibliographical citations. As noted above, some theorists have proposed that translation plagiarism can be demonstrated by the identification of common citation practices, as references are generally preserved through the disguising activity of departing from the source language. Table 2.7 presents a selection of text featuring overlap of text, and several cited references are shared between N. 2001 and Gosepath 1992.

Although the respective passages in Table 2.7 are relatively short, they each share specific bibliographical references to published works by Kippenburg, Lucheri, Winch, and Wilson. The philosopher Peter Winch is described in both passages with the same epithet ("der prominenteste Vertreter"/"the most prominent exponent") regarding cultural relativism. Furthermore, both passages assert that Winch's position is comparable to what is expressed by the philosopher Ludwig Wittgenstein. Both passages also state that there exists a debate between social scientists and philosophers on the issue of relativism, and both question the notion of a "higher" rationality

Table 2.7 Parallel bibliographical sources

N. 2001: 86–87	Gosepath 1992: 182–183
by the rise of cultural anthropology and ethnology with their investigation of myths, rites and magic in archaic societies.⁷[…the anthology edited by Hans G. Kippenberg and Brigitte Lucheri] The Occidental worldview's encounter with and discussion of mythic thinking led to the question of whether the standards of rationality in modern societies are, in fact, able to lay claim to universal validity or whether, instead, the term rational should be applied only to what is understood within the context of a particular way of life. This would mean that in modern societies a «higher» rationality of the sciences would be a chimera. The most prominent proponent of such a culturally relativist view is philosopher Peter Winch, whose position is very near that of Ludwig Wittgenstein, the initiator of this continuing debate between sociologists and philosophers (Winch 1958).⁸ [Winch developed the argument of this book in his article «Understanding a Primitive Society» (first published in *American Philosophical Quarterly* 1 1964, 307-324; reprinted in: B. R. Wilson [ed]: *Rationality*]	Das Aufkommen der Ethnologie bzw. der Kulturanthropologie und deren Untersuchungen von Mythen, Riten, Zauberei, Hexenglauben usw. führten dann zu einer radikalen Frontstellung. Mythen in archaischen Gesellschaften stehen innerhalb der uns zugänglichen kulturellen Überlieferung im schärfsten Kontrast zu dem okzidentalen Weltverständnis, das in modernen Gesellschaften herrscht. Im Spiegel des mythischen Denkens mußten deshalb die bisher nicht reflektierten Voraussetzungen des modernen Denkens sichtbar werden. Der Kontrast zwischen okzidentalem und mythischem Weltverständnis spitzte die bisherige Auseinandersetzung auf die eine fundamentale Frage zu: ob und in welcher Hinsicht die Rationalitätsstandards, von denen sich alle Sozialwissenschaftler zumindest intuitv leiten lassen, universelle Gültigkeit beanspruchen dürfen. Um diese Frage nach der Relativität der Rationalitätsstandards ist seit den sechziger Jahren zwischen Sozialwissenschaftlern und Philosophen erneut eine heftige Debatte entbrannt.⁴[Vgl. die drei, zusammengenommen recht vollständigen Sammelbände zu dieser Kontroverse B.R. Wilson (ed), *Rationality*; M. Hollis & S. Lukes (ed.), *Rationality and Relativism*; H. G. Kippenberg & B. Luchesi (Hg.).] Ausgangspunkt der z. T. sogar nach ihm benannten Debatte und sicherlich der prominenteste Vertreter des Kulturrelativismus ist Peter Winch⁵[P. Winch, *The Idea of a Social Science*, und ders., *Understanding a Primitive Society*] der unter Rückgriff auf Wittgensteins Thesen von der Unhintergehbarkeit von Sprachspielen die Auffassung vertritt, daß als rational nur das gelten kann, was man im Kontext der jeweiligen Lebensform verstehen könne […]. Damit wird es fraglich, ob wir von einem rationalen Fortschritt der Wissenschaft oder von einer »höheren« Rationalität unserer Wissenschaft ausgehen dürfen.

Table 2.8 Shared key terms in N. 2001 and Gosepath 1992

N. 2001: 93–94	Gosepath 1992: 50
opinions are defended relative to the argumentation standards possessed by the person *S* in relation to the respective facts at the time *t*. The same applies to actions which are justified relative to their underlying goals. Whenever relative arguments are present, we can speak of »rationality« in a weak sense. In contrast, a »stronger« concept of rationality would be present where the criteria themselves, to which one refers with regard to the relative substantiation of opinions and actions, once again can be proved as reasonable. The strong concept of rationality takes up the initial example from antiquity in which rationality's claim to universality was described: this accompanies the supposition that certain standards of substantiation can themselves be culturally justified. Regarding actions, this would mean that there are norms, goals, or values which themselves can be justified independent of the respective cultural province.	Meinungen werden relativ zu den Meinungen, Zielen und den Begründungsstandards begründet, die die betreffende Person zu t hat. Handlungen und Evaluationen werden relativ zu den ihnen zugrunde liegenden Meinungen und Zielen begründet, die ihrerseits rational sein müssen. Dort, wo nur relative Begründungen verlangt werden, benutzen wir »rational« in einem schwachen Sinn. Diesen Sinn nenne ich auch formale Rationalität. Es stellt sich von daher die erste Leitfrage: Gibt es daneben auch noch einen starken Rationalitätsbegriff, also das, was ich auch substantielle Rationalität nenne? Können die Kriterien, auf die wir uns zur (relativen) Begründung von Meinungen oder Handlungen beziehen, selbst nochmals als vernünftig ausgewiesen werden? Im theoretischen Bereich gäbe es möglicherweise neben dem schwachen auch einen starken Rationalitätsbegriff, dann nämlich, wenn sich zeigen ließe, daß die kulturellen Standards der Begründung (Gesetze und Theorien) ihrerseits transkulturell gerechtfertigt werden können. Diesem Problem, der Rationalität von Weltbildern, widmet sich Kapitel IV. Im praktischen Bereich erhielten wir dann einen starken Rationalitätsbegriff, wenn Normen oder Regeln (Kapitel VI) oder Ziele und Werte (Kapitel VII) ihrerseits begründet werden könnten.

in the sciences. The highly particular commonalities—bibliographic and otherwise—between these two passages suggest that something other than independent discovery is the cause of the overlap between N. 2001 and Gosepath 1992.

2.5.4 Strong and Weak Rationality

A key distinction in Gosepath 1992 is the separation of strong and weak concepts of rationality. This distinction is also found in N. 2001, as is shown in Table 2.8. Gosepath 1992's mid-book outline of later discussions of whether norms, goals, and values can themselves be justified also appears in N. 2001, but with a change. What is presented by Gosepath 1992 as an outline of topics for the latter half of the book is repackaged as definitive claim in N. 2001.

2.5.5 A Shared Thesis

According to a minority view held by some plagiarism theorists, for an act of plagiarism to occur the copied portions of an original must represent "the crux, the core, an entity, a unified or artistic whole" and "must have significant value" and (St. Onge 1988: 60). Although this view has been criticized (Dougherty 2018: 63–67), there is value to asking what portions of Gosepath 1992 are paralleled in N. 2001. Since Gosepath 1992 is a lengthy 420-page monograph, and N. 2001 is a 19-page article, one can ask whether any of the most original contributions of the monograph find expression in the shorter article. It has already been shown above that the texts in N. 2001 appear to overlap with portions of a large range of passages from Gosepath 1992. They include portions of Gosepath's Chap. 1 (Tables 2.4, 2.6), and Chap. 4

2.5 Overlap with Stefan Gosepath's *Aufgeklärtes Eigeninteresse*

Table 2.9 Rationality as "well-founded"/"wohlbegründet" in N. 2001 and Gosepath 1992

N. 2001: 92–93	Gosepath 1992: 41–43, 49–50
two general prerequisites must be fulfilled: only a being who possesses language and who is free can be considered rational. In order to give reasons, language must be presupposed. For a reasonable argument implies a conclusion reached either from the general to the specific or from the specific to the general. The rationality attributed to a person consists of the ability of setting certain facts in relation with others in a certain way. An unarticulated association, manifested only in behavior, would not be sufficient in itself. This, however, does not answer the question of whether language must be presupposed in order to have reasons. And the second prerequisite seems to me to be even more complicated; accordingly, when we call a person rational, we also attribute to him/her the ability to couple consequences to his/her reasons. This may also be best demonstrated by citing its opposite. When we accuse a person of an irrational behavior, we thereby express that this person could have decided differently had s/he taken other reasons into consideration. If a person is no longer able to be behaviorally influenced by reasons, then his/her behavior is considered compulsive. Summarizing our previous deliberations, we can note that a comprehensive meaning of rationality may be specified by the keyword «well-founded»: opinions, actions, etc, are rational when they can be justified via reasons.[40] [We cannot go into the details of the advocates of Karl Popper's falsificationism. According to this view, hypotheses cannot be conclusively proved, but only disproved. Accordingly, reason is not dependent on good reasoning, but rather on the critical role of examining and rejecting hypotheses.]	müssen zwei allgemeine Voraussetzungen für Rationalität erfüllt sein: Rational können nur Wesen sein, die Sprache besitzen und frei sind. Sprache - Wenn wir eine Person als rational bezeichnen, sprechen wir ihr die Fähigkeit zu, begründen zu können. Daß man Sprache benutzen muß, um Gründe auszudrücken, ist trivial. Die Frage ist aber, ob Sprache vorausgesetzt sein muß, um überhaupt Gründe haben zu können. Begründung impliziert einen Schluß, entweder vom Allgemeinen zum Speziellen oder vom Speziellen zum Allgemeinen. Daraus folgt, daß man sich bei der Begründung auf generelle Sachverhalte beziehen können muß. Rationalität besteht in der Fähigkeit, bestimmte gegebene Tatsachen in bestimmter angemessener Weise mit anderen zu vereinigen oder zu ihnen in Beziehung zu setzen. Eine unartikulierte, lediglich im Verhalten sich manifestierende Assoziation des speziellen Phänomens A mit einem Phänomen B reicht nicht aus […]. Wenn wir einer Person bezüglich einer bestimmten Meinung Irrationalität bescheinigen (im Gegensatz zur Arationalität), dann behaupten wir, diese Person hätte sich, wenn sie andere Gründe berücksichtigt hätte, anders entschieden bzw. anders entscheiden können. In diesem Sinn hat Rationalität etwas mit Freiheit (und Verantwortung) zu tun. Halten wir diese Person nämlich in ihrer Entscheidung für nicht mehr beeinflußbar mit Gründen, so gilt ihr Verhalten als zwanghaft und daher als nicht-rational (irrational). Aufgrund der bisherigen Analyse des Rationalitätsbegriffs läßt sich die allgemeine These formulieren: Als eine alle Anwendungssituationen umfassende Bedeutung von »Rationalität« erweist sich »wohlbegründet«. Etwas (Meinung, Handlung, Wunsch, Ziel, Norm etc,) ist rational, wenn es begründet, d. h. durch Gründe gerechtfertigt ist. Die Explikation von Rationalität im Sinne von Wohlbegründetheit wird von allen Verfechtern von K. Poppers Falsifikationismus, den sog. kritischen Rationalisten, bestritten. Nach dieser Auffassung können Hypothesen nicht gerechtfertigt werden, sondern nur widerlegt werden. Es gibt danach keine guten Gründe. Vernunft bzw. Rationalität hängt demnach auch nicht an den guten Gründen, sondern an der kritischen Rolle der Prüfung und Zurückweisung von Hypothesen.

(Tables 2.5, 2.8). As Tables 2.4, 2.5, 2.6, 2.7 and 2.8 have exhibited, the overlap between Gosepath 1992 and N. 2001 extends to many shared elements, including a meta-narrative first person claim, etymological evidence, key concepts, bibliographical references, distinctive terminology, the same variables in logical notation, and distinctive and idiosyncratic examples. These commonalities survive the translation of texts from German to English.

One may still ask further whether the central thesis of Gosepath 1992 can also be found in N. 2001. The distinctive and overarching "allgemeine These" [general thesis] of Gosepath 1992 is that the "umfassende Bedeutung" [comprehensive meaning] of "Rationalität" [rationality] can be expressed in the notion of "wohlbegründet" [well-founded] (Gosepath 1992: 49). This general thesis appears in N. 2001 in English, but again without any attribution to Gosepath (Table 2.9). The presentation of rationality as "wohlbegründet" in Gosepath 1992 occurs with a discussion of conditions of rationality and includes an account of Karl Popper's view of falsificationism, and this specific context is also repeated in N. 2001.

2.5.6 *The Downstream Literature Problem*

As shown above, several authors engaged N. 2001 in the downstream literature, and in doing so they appeared to be unknowingly dealing with Stemmer 1992 through the proxy of N. 2001. Those authors gave credit to N. 2001 for what should be credited to Stemmer 1992. When articles in the downstream literature positively discuss, cite, and quote plagiarizing articles, the original authors whose works have been misappropriated are denied recognition for their discoveries. The quality of scholarly communication is compromised and the published research literature becomes less reliable.

One can consider also how the passages in N. 2001 that overlap with Gosepath 1992 have been treated in the downstream literature. At one extreme in the downstream literature are responding articles that cite both N. 2001 and Gosepath 1992 (e.g., Tanner 2011). Evidently the identity has not been recognized by such authors. More commonly, however, downstream discussions of N. 2001 give credit to it for its views on the topic of rationality, without acknowledgment of any overlap with Gosepath 1992 (e.g., Tanner 2004) or the citations invoke N. 2001 as a general resource for the consideration of intercultural communication (Busch 2007). The downstream literature also includes positive self-citations by N. to N. 2001 for its view of rationality (e.g., N. et al. 2011). All these downstream citations—including self-citations—further embed N. 2001 in the body of published research literature.

2.6 Other European Languages

The open discussion of evidence of translation plagiarism is necessary for improving philosophical communication. The academic post-publication peer review website PubPeer contains discussions of instances of suspected translation plagiarism from the discipline of philosophy involving European languages other than German. A brief mention of them can assist in illustrating that the problem of translation plagiarism is not confined to one author of record working exclusively with German sources. There are other cases in philosophy involving different authors of record who are producing products of suspected translation plagiarism involving a variety of languages.

One recent PubPeer discussion concerns a 1988 article in English from the philosophy journal *Mind* that appears to have been largely published in French as a book chapter in 1995 under different authorship (hereafter, "E.") (PubPeer 2019). Upon being notified of the suspected plagiarism, the distributor of the volume, the publisher Libraire Droz, stopped selling the volume. Libraire Droz declined to issue a retraction, however, arguing that the editorial stewardship of the volume belonged to a now-defunct institute at University of Neuchâtel. Closer inspection shows that not only do passages from the article in *Mind* show up in English renderings in the 1995 chapter, but other sources as well. Table 2.10 presents one example of a parallel

2.6 Other European Languages

Table 2.10 English-to-French textual overlap

E. 1995: 105–106	Peacocke 1993: 1969
(a) selon la théorie comprendre la constante requiert une certaine sorte d'acceptation non inférentielle de certaines instances de P, et si (b) selon cette théorie une certaine propriété sémantique de la constante est cruciale pour la validité ou la non-validité des inférences qui la contiennent. Une théorie traite un principe comme résultant du sens d'une constante si d'une part elle ne le traite pas comme déterminant, et si d'autre part la validité du principe découle de la propriété sémantique assignée à la constante en vertu des principes qu'elle traite comme déterminant pour la constante. Par exemple nous pouvons dire que nous avons une théorie des constante de la logique propositionnelle classique qui dit que l'acceptation non inférentielle des règles classiques d'élimination et d'introduction est requise pour la compréhension du conditionnel matériel. Supposons aussi que cette théorie dise qu'une inférence propositionnelle valide est une inférence qui préserve la vérité sous toutes les assignations de valeurs de vérité aux constituants non logiques, et que ce qui fait que le conditionnel matériel exprime la fonction de vérité qu'il exprime est la fonction de vérité qui rend les règles d'introduction et d'élimination préservatrices de la vérité sous toutes les assignations. Selon cette mini théorie très élémentaire, le modus ponens, ou la règle de démonstration conditionnelle sera déterminante pour le sens, alors que la loi de Peirce ((p ⊃ q) ⊃ p) ⊃ P résulte du sens selon cette théorie.	(i) according to the theory, understanding the logical constant requires some specified kind of non-inferential acceptance of (certain) instances of the principle Π; and (ii) according to the same theory, a certain semantical property of the constant is crucial to the validity or otherwise of principles in which it occurs, [...] a theory treats a principle as s-d for a constant if [...], the validity of the principle follows from the semantical property assigned to the constant in virtue of its relations to those principles which it does treat as s-d for the constant. We can give some simple illustrations. Suppose we have a theory of the constants of classical prepositional logic which says that a certain noninferential acceptance of the classical introduction and elimination rules is required for understanding the material conditional. Suppose it says also that a valid (prepositional) argument is one which is truth-preserving under all assignments of truth-values to the non-logical constituents, and that what makes it the case that the material conditional expresses the truth-function it does is that it is the truth-function which makes the introduction and elimination rules truth-preserving under all assignments. According to this simple little theory, then, modus ponens and the rule of conditional proof will be s-d. Peirce's law ((p ⊃ q) ⊃ p) ⊃ p, on the other hand, is s-r according to this theory.

between E. (1995) from a (1993) book chapter by Christopher Peacocke.

In this case of suspected translation plagiarism from English into French, the apparent source text by Peacocke is given in a list of references at the end of the chapter. There are, however, no quotation marks or other indicators to suggest to the reader that the material has already appeared in print two years earlier, in another language, with a different author of record.

In an unrelated case, PubPeer features a discussion involving different authors of record. That discussion considers a a book chapter from a 2009 Oxford University Press collection in English that appears to have been excerpted under different authorship in Spanish in a 2015 journal article (PubPeer 2018). The apparent English source text is referenced with a bibliographical entry and some footnotes in the Spanish journal article, but without customary quotation marks that would indicate to readers the portions of the English text that are apparently being presented in Spanish translation.

2.7 Conclusion

Demonstrating cases of plagiarism can be a difficult and time-consuming task. When plagiarism involves any of the disguised varieties of plagiarism, meticulous marshalling of evidence is required to exhibit the relationship between originals and copies. It should not be surprising that acts of disguised plagiarism can take many years to be discovered. Nevertheless, in some quarters there are those who would like to impose statutes of limitation for any consideration of plagiarizing articles. In arguing against such a view, Weber-Wulff has rightly observed that it assumes the preposterous premise that "plagiarism could be 'healed' by the passage of time alone" (2014: 57). In some disciplines, particularly those in the humanities, the shelf life of a research article extends for many years if not decades, and so a research article can continue to acquire citations for a significant period after publication.

With open discussions of acts of suspected plagiarism, credit can be given to original authors whose works have been misappropriated. The cycle of continued citations to plagiarizing articles can be diminished with the public disclosure that plagiarizing articles are unreliable, and the quality of the downstream literature is thereby improved. Furthermore, open discussions of violations of scientific norms contribute to a culture where whistleblowers are encouraged. Disclosing evidence of plagiarism can be perilous, yet such activity is nevertheless essential to the enterprise of maintaining research integrity.

2.8 Postscript

In November 2018, I sent an earlier version of the text of this chapter as an article submission to *SComS*, the journal that published N. 2001. In early 2019, the manuscript was rejected without being submitted to the review process. Later in the year, the manuscript was accepted and published in the journal *Theoria* (Dougherty 2019). Shortly thereafter I wrote to the editors of *SComS* and requested a retraction of N. 2001 on the basis of (1) suspected translation plagiarism (involving the German monographs by Stemmer and Gosepath); and (2) suspected editorial misconduct (since N. was the Executive Editor of the journal at the time N. 2001 was published). In late August 2019, the editors of *SComS* retracted N. 2001. The online version of the article was removed and replaced with a brief notice that explains, "This article was retracted due to translation plagiarism" (SComS 2019: 2). The appearance of the retraction generated a short a follow-up by the editor of *Theoria* titled "Further Developments after Professor Dougherty's Recent Article." The note informed readers that the editors of *ScomS* "have now retracted the article in question due to translation plagiarism" (Hansson 2019: 344; see Weinberg 2019).

References

Busch, Dominic. 2007. *Interkulturelle Mediation*, 2nd ed. Frankfurt am Main: Peter Lang.
Danesi, Marcel, and Andrea Rocci. 2009. *Global linguistics*. Berlin: Mouton de Gruyter.
De Silva, Pali U.K, and Candace K. Vance. 2017. *Scientific scholarly communication*. Cham: Springer.
Dougherty, M.V. 2018. *Correcting the scholarly record for research integrity*. Cham: Springer.
Dougherty, M.V. 2019. The corruption of philosophical communication by translation plagiarism. *Theoria* 85 (3): 219–246.
[E.]. 1995. La signification philosophique du calcul des séquents. In *Raisonnement et calcul*, ed. Denis Miéville, 91–111. Neuchâtel: Travaux du Centre de Recherches sémiologiques de l'Université de Neuchâtel.

Elgül, Ceyda. 2019. Up to date as long as retranslated. In *Studies from a retranslation culture*, ed. Özlem Berk Albachten, and Şehnaz Tahir Gürçağlar, 117–136. Singapore: Springer.

Feix, Josef (ed. and trans.). 1980. *Herodotus Historien: Griechisch-deutsch*. Bd. 1. 3 Aufl. München: Heimeran.

Fox, Mark, and J. Jeffrey Beall. 2014. Advice for plagiarism whistleblowers. *Ethics & Behavior* 24 (5): 341–349.

Franco-Salvador, Marc, et al. 2013. Cross-language plagiarism detection using a multilingual semantic network. In *Advances in information retrieval*, ed. P. Serdyukov et al., 710–713. Berlin: Springer.

Franco-Salvador, Marc, et al. 2016. A systematic study of knowledge graph analysis for cross-language plagiarism detection. *Information Processing and Management* 52 (4): 550–570.

Gipp, Bela. 2014. *Citation-based plagiarism detection*. Wiesbaden: Springer Vieweg.

Gosepath, Stefan. 1992. *Aufgeklärtes Eigeninteresse*. Frankfurt am Main: Suhrkamp.

Hansson, Sven Ove. 2019. Further developments after Professor Dougherty's recent article. *Theoria* 85 (5): 344.

Herodotus. 1938. *Herodotus in four volumes*, vol. 2 [Trans. A. D. Godley]. London: Heinemann.

Herodotus. 1962 *The Histories of Herodotus of Halicarnassus* [Trans. Harry Carter]. Oxford: Oxford University Press.

Herodotus. 1980. *Historien: Griechisch-deutsch*. Hrsg. Josef Feix. Bd. 1. 3 Aufl. München: Heimeran.

Jianzhong, Xu. 2003. Retranslation: Necessary or unnecessary. *Babel* 49 (3): 193–202.

Kanra, Yeşim Tükel. 2019. Turkish retranslations of philosophical concepts. In *Studies from a retranslation culture*, ed. Özlem Berk Albachten Şehnaz and Tahir Gürçağlar, 41–60. Singapore: Springer.

Marcus, Adam, and Ivan Oransky. 2017. Is there a retraction problem? In *The Oxford handbook of the science of science communication*, ed. Kathleen Hall Jamieson et al., 119–126. New York: Oxford University Press.

Martin, Ben R. 2012. Does peer review work as a self-policing mechanism in preventing misconduct? In *Promoting research integrity in a global environment*, ed. Tony Mayer and Nicholas Steneck, 97–114. Toh Tuck Link: World Scientific Publishing.

Martin, Ben R. 2013. Whither research integrity? *Research Policy* 42 (5): 1005–1014.

[N.]. 2001. Rationality as a condition for intercultural understanding. *Studies in Communication Sciences* 1: 81–99. [Retracted for translation plagiarism in 2019.]

[N.], 2003. Möglichkeiten und Grenzen einer 'Ethik der Normalität des Fremden'. In *Abgrenzen oder Entgrenzen*, ed. M. Bieswanger et al., 55–63. Frankfurt am Main: IKO-Verlag.

[N.], et al. 2011. Assessing the rationality of argumentation in media discourse and public opinion. *Empedocles* 3 (1): 83–110.

Peacocke, Christopher. 1993. Proof and truth. In *Reality, representation, and projection*, ed. John Haldane and Crispin Wright, 165–190. Oxford: Oxford University Press.

PubPeer. 2018. Comment. *PubPeer*. https://pubpeer.com/publications/6A00639CE1300B91C489515D33D920.

PubPeer. 2019. Comment. *PubPeer*. https://pubpeer.com/publications/40152EAA3081FCCAC76F2962A627A8.

Rigotti, Eddo. 2003. Editors' note. *Studies in Communication Sciences* 3 (2): 5–6.

Rigotti, Eddo. 2009. *Conoscenza e significato*. Milano: Mondadori Education S.p.A.

Şahin, Mehmet, et al. 2015. Big business of plagiarism under the guise of (re)translation. *Babel* 61 (2): 192–218.

Şahin, Meymet, et al. 2019. Toward an empirical methodology for identifying plagiarism in retranslation. In *Perspectives on retranslation*, ed. Özlem Berk Albachten, and Şehnaz Tahir Gürçağlar, 166–191. New York: Routledge.

Sánchez-Vega, Fernando, et al. 2019. Paraphrase plagiarism identification with character-level features. *Pattern Analysis and Applications* 22 (2): 669–681.

SComS. 2001. [Front Matter]. *Studies in Communication Sciences* 1 (2): 1.

SComS. 2019. Retracted for translation plagiarism. https://www.e-periodica.ch/cntmng?pid=sco-003:2001:1::617.
Stemmer, Peter. 1992. *Platons Dialektik*. Berlin: Walter de Gruyter.
St. Onge, K.R. 1988. *The melancholy anatomy of plagiarism*. Lanham: University Press of America.
Tanner, Jakob. 2004. Die ökonomische Handlungstheorie vor der »kulturalistischen Wende«? In *Wirtschaftsgeschichte als Kulturgeschichte*, ed. Hartmut Berghoff and Jakob Vogel, 69–98. Frankfurt am Main: Campus.
Tanner, Jakob. 2011. »Kultur« in den Wirtschaftswissenschaften. In *Handbuch der Kulturwissenschaften*, ed. Friedrich Jaeger and Burkhard Liebsch, 195–224. Stuttgart: J.B. Metzler.
Tauginienė, Loreta, et al. 2018. *Glossary for academic integrity*. ENAI. https://www.academicintegrity.eu/wp/wp-content/uploads/2018/02/GLOSSARY_final.pdf.
Teixeira da Silva, Jaime A., and Judit Dobránszki. 2017. Notices and policies for retractions, expressions of concern, errata and corrigenda. *Science and Engineering Ethics* 23 (2): 521–554.
Turell, M. Teresa. 2008. Plagiarism. In *Dimensions of forensic linguistics*, ed. John Gibbons, and M. Teresa Turell, 265–300. Amsterdam: John Benjamins.
Weber-Wulff, Debora. 2014. *False feathers*. Heidelberg: Springer.
Weinberg, Justin. 2019. Translation plagiarism in philosophy. *Daily Nous*. http://dailynous.com/2019/10/01/translation-plagiarism-philosophy. Accessed 1 Oct 2019.
Zhang, Yuehong (Helen). 2016. *Against plagiarism*. Cham: Springer.

Chapter 3
Compression Plagiarism

Abstract Despite an increased recognition that plagiarism in published research can take many forms, current typologies of plagiarism are far from complete. One under-recognized variety of plagiarism—designated here as *compression plagiarism*—consists of the distillation of a lengthy scholarly text into a short one, followed by the publication of the short one under a new name with inadequate credit to the original author. In typical cases, compression plagiarism is invisible to unsuspecting readers and immune to text-matching software. The persistence of uncorrected instances of plagiarism in all its forms—including compression plagiarism—in the body of published research literature has deleterious consequences for the reliability of scholarly communication. Not the least of these problems is that original authors are denied credit for their discoveries. When unsuspecting researchers read articles that are the products of plagiarism, they unwittingly engage the arguments of hidden original authors through the proxy of plagiarists. Furthermore, when these researchers later publish responses to the plagiarizing articles, not knowing they are engaging products of plagiarism, they create additional inefficiencies and redundancies in the body of published research. This chapter considers the particular ways in which compression plagiarism weakens the quality of scholarly argumentation, with special attention paid to the field of philosophy.

Keywords Compression plagiarism · Authorship · Research misconduct · Retractions · Argumentation · Scholarly communication

Compression plagiarism can be understood as a distinct variety of disguised plagiarism. It occurs when a lengthy published text is distilled or condensed into a short text, and this short text is published under a new name without adequate reference to the lengthy original. The lack of proper attribution in the plagiarizing short text to its source causes readers to remain ignorant that the short text originates elsewhere. Furthermore, the concentration of the lengthy text into a plagiarizing shorter one renders it largely immune to the forensic work of text-matching software.

Compression plagiarism is a form of disguised plagiarism because the compression of the source text obscures the plagiarizing text's dependency on its source. The compression may be twofold. First, the compression may occur at the macro level and involve the appropriation of passages that are very distant in the source text. For example, a plagiarist might take portions from the introduction, central chapters, and conclusions of a lengthy book in producing a very short journal article. Second, the compression may occur at the micro level and involve the reduction of a paragraph into a sentence (or a long sentence into a shorter one).

Identifying cases of compression plagiarism is not easy because the genetic or derivative relationship of the short text from the lengthy one is hidden. Acts of compression plagiarism are not obvious like the copy-and-paste variety; they are much less likely to be flagged as products of plagiarism, and editors are less likely ever to issue retractions for them. Nevertheless, they mar the reliability of the body of published research no less than the more manifest forms of plagiarism.

3.1 A Case of Compression Plagiarism

To explore the problem of compression plagiarism, I examine here one case that appeared as an article in the philosophy journal *Argumentation*. My aim is to illustrate the kinds of evidence that are relevant for a demonstrating this variety of disguised plagiarism and to illustrate how to present the evidence. The analysis that follows can be viewed as an exercise in post-publication peer review (PPPR) as well as an exhibition regarding how to prove plagiarism. The article in question appeared in the journal as part of a themed issue on strategic maneuvering in 2006. The editors' introduction explains that the articles in the issue were first presented as papers at a 2006 conference in Amsterdam funded by The Netherlands Organisation for Scientific Research and that all the main papers and the shorter commentary papers are published in the issue in "amended" form (van Eemeren and Houtlosser 2006: 378). The author of record for the article in question (hereafter, "N.") produced one of the commentary papers (N. 2006), and it addresses a lengthy paper by Christopher W. Tindale that was titled "Constrained Maneuvering: Rhetoric as a Rational Enterprise" (Tindale 2006). N.'s short commentary paper, which is just over four pages, divides into two parts. The first half summarizes Tindale's article with nine documented quotations. The remaining half—two pages or so—provides what appears to be an original contribution as it purports to advance an understanding of rationality by distinguishing a strong from a weak concept of rationality (or reasonableness). This latter half of the article gives the appearance of being original interpretive work, as it is presented with meta-narrative language in which N. is speaking in the first person (e.g., "I shall further demonstrate […]"; "I mean […]"; "In order to clarify my point […]") (N. 2006: 469).

Much of content of that half of the article, however, appears to bear a relationship to a previously published work that is nowhere referenced in the short 2006 article. That is, there appears to be overlap between the interpretive core part of the short 2006 article and a 420-page monograph published in German 14 years earlier by philosopher Stefan Gosepath. Titled *Aufgeklärtes Eigeninteresse: Eine Theorie theoretischer und praktischer Rationalität* [Enlightened Self-Interest: A Theory of Theoretical and Practical Rationality] (Gosepath 1992), the lengthy apparent source

3.1 A Case of Compression Plagiarism

text is nowhere cited in any way in the 2006 article. Tables 3.2, 3.3, 3.4, 3.5, 3.6, 3.7 and 3.8 in the Appendix below provide seven select passages from the short 2006 article, in the order as they appear the journal, exhibiting texts that overlap with Gosepath's (1992) monograph. The highlighting identifies the commonalities between the two works.

Most of the printed lines of the last half of N.'s article have some portion overlapping with text in Gosepath's book. The texts as they appear in the 2006 article are "compressed" in the two senses described earlier. First, they are compressed at the macro level insofar as the 2006 article appears to be constructed from passages taken from a lengthy span of Gosepath's large book. The examples in the Appendix (Tables 3.2, 3.3, 3.4, 3.5, 3.6, 3.7 and 3.8) exhibit a selection of passages constituting the 2006 article that overlap with passages taken from pages 22–24, 49–50, 184–185, and 193 of Gosepath (1992). They include passages from the beginning of the first chapter of Gosepath's monograph (Table 3.2), up through the end of part two (Table 3.8), including Gosepath's mid-book outline about what will follow in the later parts of the book (Table 3.4). Additionally, the texts are compressed at the micro level insofar as they often appear as abbreviations of Gosepath's work: the individual passages are longer in Gosepath's monograph and appear to be contracted in N.'s article. Furthermore, the order in which the texts appear in N. 2006 generally follows the order in which they appear in Gosepath (1992), which further strengthens the case of dependency of the later article on the earlier book.

The textual parallels between Gosepath (1992) and N. 2006 excerpted in Tables 3.2, 3.3, 3.4, 3.5, 3.6, 3.7 and 3.8 concern highly specific and particularized claims about rationality. The ostensibly most significant claim put forth about the concept of rationality in the 2006 article is the distinction between strong and weak concepts of rationality or reasonableness, and this distinction is introduced and defended at length in Gosepath's book and is presented as the distinction between a strong (starken) and a weak sense (schwachen Sinn) of rationality (Rationalität). The overlap of presentation of this key distinction is striking. For example, both texts make the following identical claim:

> Whenever relative arguments are present, we can speak of "reasonableness" in a weak sense. (N. 2006: 469)

> Dort, wo nur relative Begründungen verlangt werden, benutzen wir » rational « in einem schwachen Sinn. (Gosepath 1992: 50)

Gosepath also explains that his "allgemeine These" [general thesis] is that the "umfassende Bedeutung" [comprehensive meaning] of "Rationalität" [rationality] can be expressed in the notion of "wohlbegründet" [well-founded] (Gosepath 1992: 49). N. 2006 also puts forward the notion of "well-founded" as the key to understanding the various applications of rationality:

> "well-founded": opinions, actions, etc. are reasonable when they can be justified via reasons. (N. 2006: 469)

> »wohlbegründet « . Etwas (Meinung, Handlung, Wunsch, Ziel, Norm etc.) ist rational, wenn es begründet, d. h. durch Gründe gerechtfertigt ist. (Gosepath 1992: 50)

The overlap of N. 2006 and Gosepath (1992) is not restricted to the presentation of identical notions using the same words, however. When it comes to setting up a framework for discussing the rationality of opinions, the variables and nomenclature used for formalizing opinions are almost identical in both texts:

> So, holding an opinion is a relation between a subject (S) and a proposition… at a certain point in time or during a certain period of time (t), i.e. 'S is of the opinion p at time t.' (N. 2006: 469)

> Das Meinen ist eine Relation zwischen einer Person und einer Proposition zu einem Zeitpunkt oder in einem Zeitraum: S meint zu t, daß p. (Gosepath 1992: 24)

The overlap between N. 2006 and Gosepath (1992) also includes the definition of a key term and the use of the first person to introduce this stipulative definition:

> By cognitive condition, I mean the whole set of accepted opinions, convictions, evidences, and goals possessed by a certain person at a certain period in time […]. (N. 2006: 470)

> Unter einem kognitiven Ausgangszustand fasse ich den Korpus der akzeptierten Meinungen, die zur Verfügung stehenden Evidenzen und die Ziele, die eine bestimmte Person zu einem bestimmten Zeitpunkt hat. (Gosepath 1992: 470)

Several distinctions pertaining to rationality and its synonyms as identified by Gosepath (e.g., reasonableness) are also found in both N. 2006 and Gosepath 1992 (see Tables 3.5 and 3.7).

In light of these points, it cannot be argued that the texts that appear to be shared between N. 2006 and Gosepath 1992 are minor or non-essential portions of Gosepath's monograph, but they pertain to the very core of Gosepath's project. That is, the overlap does not consist of commonly used terms found in communication and argumentation studies, but rather the parallels concern highly idiosyncratic philosophical writing, and the identity is apparent even as the text moves from the original German to a formulation in English.

Examples of compression plagiarism in the published research literature are very difficult to identify. This does not mean, necessarily, that compression plagiarism is extremely rare in the published research literature. As noted above, compression plagiarism is typically immune to text-matching software. In the plagiarizing version, the compressed text segments are no longer sufficiently adjacent to allow them to be flagged as overlapping with the source text. It is very difficult for unsuspecting readers to identify compression plagiarism in the course of ordinary research. Even were a reader to be on the lookout for subtle plagiarism of this kind, the compression of a large text into a small one can impede a reader from recognizing an identity between texts that are so different in size. If passages that are dispersed over hundreds of pages in the source work appear in a couple of pages in their plagiarizing versions, then the degree of compression virtually guarantees that the plagiarism will escape detection. In some cases, many years—even decades—may pass before the plagiarism is discovered.

When a single act of compression plagiarism involving a very large text and a very small text is combined with some other form of plagiarism, then the odds that it will be discovered become even less likely. For instance, if an act of compression plagiarism is also a case of translation plagiarism, then most who encounter the plagiarizing text will likely never suspect that it could be a case of plagiarism. In her typology of plagiarism, Weber-Wulff observes that in translation plagiarism "the plagiarist chooses a text portion in a language different from the target language," and that if the translation quality is high it will more easily escape detection (7). (For more on translation plagiarism, see Chap. 2 in this volume.)

N. 2006 is not only a case of compression plagiarism but is also a case of translation plagiarism: the apparent source text is in the German language, and the 2006 article is in English. Acts of compression plagiarism that are then modified a second time as acts of translation plagiarism create an almost insurmountable distance between original and copy that is unlikely to be overcome by even the most careful of readers, including journal editors, peer reviewers, any co-authors, a plagiarist's research colleagues, others in the field, and authorities at the home institution of a plagiarist.

3.2 How Compression Plagiarism Corrupts Scholarly Communication

Theorists who consider plagiarism in published scholarly research typically focus on two kinds of harm that plagiarism causes. Plagiarism harms the original authors who are denied credit for their discoveries, and plagiarism harms the larger system of knowledge (Gipp 2014: 9). The time and resources of peer reviewers, journal editors, and readers are wasted as plagiarizing articles produce duplicated research articles that take up valuable space in journals that should have been reserved for genuine products of research. Furthermore, on the basis of plagiarizing articles, an academic plagiarist creates the illusion of research productivity, and this façade may lead to undeserved grants, promotions, and employment opportunities that should have been bestowed on authentic researchers.

Such considerations of the harm that plagiarism causes are accurate, but there is additional harm that is largely unrecognized: plagiarism severely impacts the quality of scholarly argumentation. Weber-Wulff rightly observes that "when academics plagiarize they are damaging academic discourse" (2014: 22). How, exactly, such damage occurs and how it affects the quality of academic discourse is not always clearly expressed. To illustrate the devastating effects of compression plagiarism on scholarly communication one may consider N. 2006 as a test case. If N. 2006 plagiarizes Gosepath 1992, the following unpleasant claims are true:

(i) Gosepath, the original author, is denied credit for his 1992 original work when it re-appears in print in condensed form under N.'s name in the 2006 article.
(ii) Gosepath's work has a double representation in the body of published research literature: first as Gosepath 1992 and then as N. 2006.
(iii) Readers of N.'s 2006 article are misled about the source of the argumentation it presents.

(iv) What appears to be an original short commentary by N. on Tindale's article is really a positioning of Gosepath's book in relation to Tindale's article.
(v) The real interlocutor with Tindale is not N., but Gosepath, whose work has been conscripted into plagiarizing service.

Claims (i) and (ii) express that plagiarizing scholarly articles deny credit to genuine authors for their discoveries and generate duplications in the body of published research literature. Claims (iii)–(v) identify elements of the breakdown of scholarly communication. Presumably Tindale and readers of N.'s 2006 article do not know that they are engaging Gosepath's work through the proxy of N.

There are two paths, therefore, to Gosepath's work in the body of published research literature. In the first path, readers encounter the book directly, under Gosepath's name, in the 1992 monograph itself. In the second path, readers encounter Gosepath's book in a hidden or disguised way in N. 2006. The first path is a normative one, and the second is an aberration, insofar as N. 2006 is a proxy through which readers unknowingly encounter the content of Gosepath (1992). A reader who arrives at the content of Gosepath (1992) through the proxy of N. 2006 is not doing so efficiently.

The harm to scholarly communication is not limited to the abovementioned claims, however. In the published version of Tindale (2006), Tindale addresses N.'s commentary, but without giving any indication of any awareness that N.'s short article overlaps with Gosepath's book. Quite the opposite appears to be the case. Tindale states:

> In his comments, [N.] rightly identifies the challenge here to be one regarding the relationship between the particular cognitive claims and the standard provided by the universal audience. [N.] points out that if an opinion is to be held in a rational way, then conditions related to both of these must be met. On this I think we would be in agreement. (Tindale 2006: 464)

Elsewhere in the article, Tindale writes, "I would like to thank my Amsterdam commentator, [N.], for his constructive critique" (464). In light of these remarks, therefore, to (i)–(v) can be added:

(vi) When Tindale later responds to N. in print, he is (unknowingly) responding to Gosepath through the proxy of N.

Additional evidence in support of (vi) is Tindale's continued acknowledgment of N.'s contributions to the debate in a revised version of Tindale (2006) that was published a few years later in a collection of essays on the topic of strategic maneuvering (Tindale 2009: 59). One can surmise that the quality of argumentation in an academic debate must be compromised if the main participant is not aware of the true identify of his principal interlocutor.

One might further consider that when a lengthy monograph is radically abbreviated in an act of compression plagiarism that reduces it to only a couple of pages, then the intelligibility of the plagiarizing article must be doubtful. The meaning of the abbreviated selections, when they appear in the plagiarizing article, are no longer embedded in their proper context. If meaning is largely context-dependent, how can severe compression plagiarism generate an intelligible article? In this particular case

3.2 How Compression Plagiarism Corrupts Scholarly Communication

of compression plagiarism, however, there is some evidence that this objection fails. The subsequent scholarly literature looks quite favorably at the interchange between Tindale (2006) and N. 2006. One researcher praises N. 2006 in print, claiming, "for a particularly good example, see [N.]'s treatment of *reasonableness* in his response to Christopher Tindale 2006, p. 469" (Fahnestock 2009: p. 197). To the mind of that researcher, N.'s contribution is a valuable one, so one may be tempted to conclude that this instance of compression plagiarism has succeeded in producing something not only judged to be intelligible, but enlightening, by experts in the field. Therefore, to (i)–(vi) can be added:

(vii) When other researchers positively cite N. 2006, they unwittingly credit to N. what should be credited to Gosepath.

The approach by Fahnestock 2009 is not an outlier. More recently, the interchange between Tindale (2006) and N. 2006 has been briefly summarized in a magisterial 2014 compendium, *Handbook of Argumentation Theory*, and as part of its discussion it includes a 4-line extract quotation from N. 2006. The quotation happens to be part of the text identified here in Table 3.6, which appears to overlap with text in Gosepath's book (van Eemeren et al. 2014: 416). The fact that what these commentators highlight as distinctive and original to N. 2006 is really found in Gosepath 1992 confirms that the overlap pertains to idiosyncratic significant content from Gosepath's book, rather than that common expressions found in the writings of many authors.

The continued positive citations to N. 2006 in later literature, with no indication of awareness of the apparent overlap between N. 2006 and Gosepath (1992), further embeds the 2006 article in the published research literature. By commending N. 2006 to additional readers, these responding articles perpetuate content from Gosepath (1992), but under the wrong authorship, crediting N. rather than Gosepath. The quality of scholarly argumentation is severely impacted as more researchers appear to encounter Gosepath (1992) under the guise of N. 2006, without knowing that it originates in Gosepath (1992).

The scholarly audience for this act of compression plagiarism has grown significantly since 2006. It appears to include no less than: Tindale, the original conference attendees, the pre-publication journal peer reviewers for N. 2006, the readers of the published version of N. 2006, those who produce responding articles to N. 2006, and all those who encounter N. 2006 as cited and extracted in the responding articles. If what I have proposed is correct, there are three philosophers involved in this scholarly exchange in the published research literature: Tindale, N., and Gosepath, but to most observers Gosepath's apparently conscripted participation is not evident. In light of (i)–(vii), it should be concluded that this case of compression plagiarism has resulted in a wide-ranging corruption of scholarly communication.

3.3 Objections

There may be some resistance in some quarters to recognizing a general typology of plagiarism that includes various disguised forms of plagiarism, such as compression plagiarism or translation plagiarism. Yet, if editors and publishers restrict the issuance of retractions to more manifest forms of plagiarism (e.g., copy-and-paste) that are discovered, then the research literature will persist in being unreliable in many instances. That is, if retractions are limited to obvious plagiarism cases, then many unreliable plagiarizing articles will remain as part of the body of published research literature to the detriment of researchers. The issuance of retractions for all forms of plagiarism, including its subtler disguised forms, will insure the most reliable body of published research literature for researchers.

Perhaps one might grant in principle the necessity for editors and publishers to retract acts of compression plagiarism yet remain unsure whether the evidence of compression plagiarism offered here concerning N. 2006 is sufficiently dispositive to warrant retraction in this case. That is, one might wonder whether the evidence sufficiently establishes that Gosepath's 420-page monograph is the source for the second half of the very short article. Could the overlap merely be accidental or fortuitous? To underscore further the unlikeliness of that position, I offer here three additional considerations, in increasing order of strength, in support of the view that the overlap between N. 2006 and Gosepath (1992) is not accidental.

First, one might consider that 13 other articles by N. have recently received corrections for plagiarism and failures to credit sources correctly (see Illarietti 2018; Weinberg 2018a, b; Stern 2018a, b; Palus 2016a, b; Dougherty 2018: 153–189). Those 6 retractions, 3 errata, and 4 corrigenda include the two corrections that appeared in *Argumentation* in 2015 for other works by N. (van Eemeren 2015a, b). Nevertheless, even though academic plagiarism is generally serial (Fox and Beall 2014: 346), the mere fact that there are defects in other articles by N. does not in itself establish any defects in the particular case of N. 2006.

Second, one might consider whether it can be determined if N. was familiar with Gosepath (1992) prior to the publication of N. 2006. There appears to be no linguistic barrier to having encountered Gosepath's work, since many of the publications for which N. is the author of record are in German. Incontrovertible public evidence that N. was aware of Gosepath (1992) prior to the publication of N. 2006 is that N. cites Gosepath's monograph in earlier published work. This fact demonstrates that N. was familiar to some degree with Gosepath's monograph. Although Gosepath's monograph is nowhere cited or referenced in the N. 2006 article in *Argumentation*, N. does provide a bibliographical reference to Gosepath in a work published five years earlier (N. 2001).

Third, and most importantly, an examination of text of N. 2001 reveals an important clue: almost the entire second half of N. 2006—the portion of the article that purports to be original work—is found to be already published in N. 2001 with only minor changes. That is, nearly all the sentences and paragraphs of the second half N. 2006 published in *Argumentation* in 2006 can be found in N. 2001 published in

3.3 Objections

Table 3.1 A display of passages from N. 2001 and N. 2006 in relation to Gosepath 1992

N. 2001 and Gosepath 1992		N. 2006 and Gosepath 1992	
N. 2001: 93–94	Gosepath 1992: 50	N. 2006: 469–470	Gosepath 1992: 50
opinions are defended relative to the argumentation standards possessed by the person S in relation to the respective facts at the time *t*. The same applies to actions which are justified relative to their underlying goals. Whenever relative arguments are present, we can speak of »rationality« in a weak sense. In contrast, a »stronger« concept of rationality would be present where the criteria themselves, to which one refers with regard to the relative substantiation of opinions and actions, once again can be proved as reasonable. The strong concept of rationality takes up the initial example from antiquity in which rationality's claim to universality was described: this accompanies the supposition that certain standards of substantiation can be culturally justified. Regarding actions, this would mean that there are norms, goals, or values which themselves can be justified independent of the respective cultural province.	Meinungen werden relativ zu den Meinungen, Zielen und den Begründungsstandards begründet, die die betreffende Person zu t hat. Handlungen und Evaluationen werden relativ zu den ihnen zugrunde liegenden Meinungen und Zielen begründet, die ihrerseits rational sein müssen. Dort, wo nur relative Begründungen verlangt werden, benutzen wir »rational« in einem schwachen Sinn. Diesen Sinn nenne ich auch formale Rationalität. Es stellt sich von daher die erste Leitfrage: Gibt es daneben auch noch einen starken Rationalitätsbegriff, also das, was ich auch substantielle Rationalität nenne? Können die Kriterien, auf die wir uns zur (relativen) Begründung von Meinungen oder Handlungen beziehen, selbst nochmals als vernünftig ausgewiesen werden? Im theoretischen Bereich gäbe es möglicherweise neben dem schwachen auch einen starken Rationalitätsbegriff, dann nämlich, wenn sich zeigen ließe, daß die kulturellen Standards der Begründung (Gesetze und Theorien) ihrerseits transkulturell gerechtfertigt werden können. Diesem Problem, der Rationalität von Weltbildern, widmet sich Kapitel IV. Im praktischen Bereich erhielten wir dann einen starken Rationalitätsbegriff, wenn Normen oder Regeln (Kapitel VI) oder Ziele und Werte (Kapitel VII) ihrerseits begründet werden könnten.	Whenever relative arguments are present, we can speak of "reasonableness" in a weak sense. In contrast, a "strong" concept of rationality requires that the criteria to which one refers in the process of the relative substantiation of opinions and actions can themselves be proved as reasonable. The strong concept of reasonableness lays thus claim to universality: it implies that certain standards of substantiation can be justified independently of any audience. Regarding arguments, this would mean that there are norms, goals, or values which can be justified independently of a given specific audience which is being addressed.	Meinungen werden relativ zu den Meinungen, Zielen und den Begründungsstandards begründet, die die betreffende Person zu t hat. Handlungen und Evaluationen werden relativ zu den ihnen zugrunde liegenden Meinungen und Zielen begründet, die ihrerseits rational sein müssen. Dort, wo nur relative Begründungen verlangt werden, benutzen wir »rational« in einem schwachen Sinn. Diesen Sinn nenne ich auch formale Rationalität. Es stellt sich von daher die erste Leitfrage: Gibt es daneben auch noch einen starken Rationalitätsbegriff, also das, was ich auch substantielle Rationalität nenne? Können die Kriterien, auf die wir uns zur (relativen) Begründung von Meinungen oder Handlungen beziehen, selbst nochmals als vernünftig ausgewiesen werden? Im theoretischen Bereich gäbe es möglicherweise neben dem schwachen auch einen starken Rationalitätsbegriff, dann nämlich, wenn sich zeigen ließe, daß die kulturellen Standards der Begründung (Gesetze und Theorien) ihrerseits transkulturell gerechtfertigt werden können. Diesem Problem, der Rationalität von Weltbildern, widmet sich Kapitel IV. Im praktischen Bereich erhielten wir dann einen starken Rationalitätsbegriff, wenn Normen oder Regeln (Kapitel VI) oder Ziele und Werte (Kapitel VII) ihrerseits begründet werden könnten.

Studies in Communication Sciences, strewn over pages 88–89, 93, and 94–95. When N. 2006 appeared in print in *Argumentation*, it constituted an instance of substantial undocumented text recycling from N. 2001. That acts of text recycling exist is not in itself surprising, although editors and publishers are becoming increasingly less tolerant of the practice in some fields (Horbach and Halffman 2019; Pemberton et al. 2019). What is most important here, however, is that the text common to both N. 2001 and N. 2006 overlap with Gosepath (1992), and in the earlier text—N. 2001—*the overlap with Gosepath is even more evident.*

When one compares how the compression plagiarism in N. 2006 fares when set aside the same passages as they appear in 2001, one will see that there is even more overlap with Gosepath (1992). Table 3.1 offers an example, and the additional overlap is indicated in bold, and all overlap is indicated in highlighting.

Table 3.1 reveals that N. 2001 has even more overlap with Gosepath (1992) than does N. 2006. The additional texts—before and after and highlighted in bold—further demonstrate an apparent dependency of N.'s texts on Gosepath (1992). This point suggests that the three texts are related to each other. To put the matter another way: N. 2006 appears to be an instance of compression plagiarism that appropriates Gosepath (1992), and the argument for dependency is strengthened further when one sees the relationship between the two in light of an intermediate text which is N. 2001.

To illustrate this variety of disguised plagiarism, I have contended that N. 2006 is a case of compression plagiarism. The primary evidence that I have offered is textual: N. 2006 contains highly idiosyncratic text parallels with Gosepath (1992) that include the distinction between strong and weak rationality, the concept of rationality as "well-founded," and as well as distinctive definitions, precisions, and nomenclature. This primary textual evidence was supplemented, however, with additional evidence, namely, that N. cites Gosepath (1992) in work prior to N. 2006, thus countering any suggestion that Gosepath 1992 and N. 2006 could represent the simultaneous discovery of the same positions, in different languages, 14 years apart. A consideration of this circumstantial evidence revealed, however, another piece of confirmatory textual evidence: N. 2006 substantially consists of undocumented recycled text from N. 2001, and the overlap with Gosepath (1992) is even more manifest in N. 2001. When a passage recycled in N. 2006 was viewed as it exists in N. 2001, even more overlap with Gosepath is found, as text before and after the passage was seen to overlap with Gosepath (1992).

3.4 Conclusion

In this chapter I have argued that compression plagiarism constitutes a distinct variety of plagiarism.[1] As a disguised form of plagiarism, it is typically not evident to most readers nor is it identifiable through standard text-matching software. Compression plagiarism is especially damaging to the quality of scholarly argumentation for several reasons. Readers unwittingly encounter the arguments of others through the proxy of plagiarists, and this phenomenon engenders errors about whom the principal interlocutors in academic debates truly are. Not only are genuine authors denied credit for their work when plagiarizing articles are published, but the arguments of the genuine authors continue to be attributed to others as the plagiarizing articles are discussed, cited, and quoted in the downstream literature in corrupted responding

---

[1] For additional examples of compression plagiarism, see Chaps. 5.2.1 and 6.5.

3.4 Conclusion

articles. Compression plagiarism therefore has pernicious and enduring effects on the quality of scholarly communication.

When faced with demonstrable evidence of plagiarizing articles, editors and publishers should issue clear statements of retraction. These retraction statements should be explanatory to correct the scholarly record fully. Credit should be given to the original sources and the retractions should be published within the pages of the affected journals. Such actions strengthen the relationship a journal has to its readers, minimize the chances of reprisals against whistleblowers, and encourage vigilance in ensuring the quality of scholarly argumentation.

3.5 Postscript

An earlier version of this chapter was published in the journal *Argumentation* in April 2019. In that version, the defects in N. 2006 were characterized as "a suspected instance of compression plagiarism" (Dougherty 2019: 391). Five months after publication, however, a brief correction by N. appeared in *Argumentation*. It stated in part:

> English translations of phrasings regarding several distinctions in the concept of rationality were taken from Stefan Gosepath's book *Aufgeklärtes Eigeninteresse: Eine Theorie theoretischer und praktischer Rationalität*, Frankfurt: Suhrkamp 1992, without providing reference to his publication. (N. 2020: 285)

The correction concluded with a conditional expression of regret and apology: "I express regret if I have impinged upon the author's rights and apologize for this error of omission" (N. 2020: 285). Although it did not use the term *compression plagiarism*, the correction concedes the dependency of N. 2006 on Gosepath 1992.

The identification of the phenomenon of compression plagiarism, first occasioned by the consideration of N. 2006, has been beneficial beyond generating the correction of that short article. The term is now established as precise term in discussions of research integrity. The nomenclature has been noted on the website *Retraction Watch* (Oransky 2019). Furthermore, the term has been adopted by the publisher Brill in its new statement of research ethics policies. The section on plagiarism now has an entry titled "Compression plagiarism" which is defined as "Distillation and repurposing of the words, ideas, methods, results, or artwork of a substantially longer work without appropriate citation" (Brill 2019: 5).

Appendix

See Tables 3.2, 3.3, 3.4, 3.5, 3.6, 3.7 and 3.8.

Table 3.2 Compression (a)

N. 2006: 469	Gosepath 1992: 23–24
on the objective area to which it is applied: deliberation, opinions, or arguments may be candidates for reasonableness within the cognitive area; actions within the behavioral area; and desires within the emotional area. When we call discourses, arguments, or thoughts reasonable, we are referring to them neither as to linguistic constructs nor as regards the respective content of their declarative statements (technically speaking, the proposition or the propositional subject matter). Such a content is either true or false, but not reasonable or unreasonable. Rather, by using the expression "reasonable" we are describing opinions in a much broader sense. In the following, "opinion" should be understood as a proposition held to be true. Holding a proposition to be true includes believing, expecting, supposing something, being convinced *of something*, considering its possibility, etc. So, holding an opinion is a relation between a subject (S) and a proposition (P), which can be formalized as "*S* is of the opinion that *p*...". To be exact, one has to add: S is of the opinion *p* at a certain point in time or during a certain period of time (*t*), i.e. "*S* is of the opinion *p* at time *t*."	Im theoretischen Bereich, verstanden als die kognitive Sphäre, scheinen sich auf den ersten Blick eine Menge Kandidaten für Rationalität anzubieten. Beispiele dafür wären: Überlegungen, Überzeugungen, Gedanken, Reden, Argumente usw. Was ist es, was wir in diesem Bereich als rational bezeichnen? Offenbar ist nicht die Äußerung als sprachliches Gebilde rational oder irrational. Denn, um zu beurteilen, ob es vernünftig ist, einen bestimmten Behauptungssatz in einer Situation zu äußern, müssen zusätzliche Situationsumstände mitberücksichtigt werden. Aber auch der Inhalt eines Aussagesatzes, die Proposition, kann weder rational noch irrational sein, sie ist wahr oder falsch. Als rational oder irrational bezeichnen wir umgangssprachlich vielmehr Meinungen. Unter »Meinung« soll hier das Fürwahrhalten einer Proposition verstanden werden.[4] [Hier geht es um das Problem, wie »eine Meinung haben« als mentaler Zustand philosophisch genau zu charakterisieren ist.] Wir benutzen dafür Worte wie: glauben, erwarten, annehmen, überzeugt sein, für möglich halten, für wahrscheinlich halten, als unmöglich ansehen, ahnen, auf etwas gefaßt sein, sich etwas einbilden, vermuten, mit etwas rechnen etc. Das ist eventuell mehr, als wir umgangssprachlich unter dem Begriff einer Meinung verstehen. Rational oder irrational ist es, eine Meinung zu haben, nicht der Inhalt der Meinung, die Proposition. Das Meinen ist eine Relation zwischen einer Person und einer Proposition zu einem Zeitpunkt oder in einem Zeitraum: S meint zu t, daß p.

Table 3.3 Compression (b)

N. 2006: 469	Gosepath 1992: 22, 49–50
In contrast with simple descriptive words, the expression "unreasonable" possess a normative component in our cultural context. To say that someone is behaving unreasonably means not only to suggest that a certain statement or action of the subject has the named characteristic. It also usually includes a negative evaluation or criticism of such a statement or action, since standards of reasonability are not fulfilled. These standards may be rendered briefly by the keyword "well-founded": opinions, actions, etc. are reasonable when they can be justified via reasons. Hence it is obvious that the meaning of "reasonable" is relative: opinions are defended relative to the argumentation standards possessed by the subject *S* in relation to the respective facts at the time *t*.	die Ausdrücke »vernünftig« und »rational« außer einer deskriptiven auch eine normative Bedeutungskomponente besitzen. Zu sagen, etwas sei irrational, heißt, es negativ zu bewerten und so zu kritisieren, weil es bestimmte Standards der Vernunft oder Rationalität nicht erfüllt. [...] Als eine ist Anwendungssituationen umfassende Bedeutung von »Rationalität« erweist sich »wohlbegründet«. Etwas (Meinung, Handlung, Wunsch, Ziel, Norm etc.) ist rational, wenn es begründet, d. h. durch Gründe gerechtfertigt ist. Zwie Problemkreise werden den Fortgang der Untersuchung bestimmen: Auf einer ersten Ebene handelt es sich bei der Rationalität immer um eine relative Begründung. Meinungen werden relativ zu den Meinungen, Zielen und den Begründungsstandards begründet, die die betreffende Person zu t hat.

Table 3.4 Compression (c)

N. 2006: 469–470	Gosepath 1992: 50
Whenever relative arguments are present, we can speak of "reasonableness" in a weak sense. In contrast, a "strong" concept of rationality requires that the criteria to which one refers in the process of the relative substantiation of opinions and actions can themselves be proved as reasonable. The strong concept of reasonableness thus claims to universality: it implies that certain standards of substantiation can be justified independently of any audience. Regarding arguments, this would mean that there are norms, goals, or values which can be justified independently of a given specific audience which is being addressed.	Dort, wo nur relative Begründungen verlangt werden, benutzen wir »rational« in einem schwachen Sinn. Diesen Sinn nenne ich auch formale Rationalität. Es stellt sich von daher die erste Leitfrage: Gibt es daneben auch noch einen starken Rationalitätsbegriff, also das, was ich auch substantielle Rationalität nenne? Können die Kriterien, auf die wir uns zur (relativen) Begründung von Meinungen oder Handlungen beziehen, selbst nochmals als vernünftig ausgewiesen werden? Im theoretischen Bereich gäbe es möglicherweise neben dem schwachen auch einen starken Rationalitätsbegriff, dann nämlich, wenn sich zeigen ließe, daß die kulturellen Standards der Begründung (Gesetze und Theorien) ihrerseits transkulturell gerechtfertigt werden können. Diesem Problem, der Rationalität von Weltbildern, widmet ich Kapitel IV. Im praktischen Bereich erhielten wir dann einen starken Rationalitätsbegriff, wenn Normen oder Regeln (Kapitel VI) oder Ziele und Werte (Kapitel VII) ihrerseits begründet werden könnten.

Table 3.5 Compression (d)

N. 2006: 470	Gosepath 1992: 184
At this point it seems necessary to introduce another distinction not always sufficiently addressed in the discussion of the relative validity of rationality. Whether or not it is reasonable to have opinion p depends on two conditions: on the one hand, on the respective cognitive condition; on the other hand, on the rules of rationality.	In welchen Hinsichten ist der bisher ermittelte Sinn von »rational« relativ? Ob es rational ist, p zu glauben, ist abhängig von einem kognitiven Ausgangszustand, von dem aus entsprechend den Regeln der Rationalität die Meinung, daß p, gebildet wurde.

Table 3.6 Compression (e)

N. 2006: 470	Gosepath 1992: 185
By cognitive condition, I mean the whole set of accepted opinions, convictions, evidences, and goals possessed by a certain person at a certain period in time	Unter einem kognitiven Ausgangszustand fasse ich den Korpus der akzeptierten Meinungen, die zur Verfügung stehenden Evidenzen und die Ziele, die eine bestimmte Person zu einem bestimmten Zeitpunkt hat.

Table 3.7 Compression (f)

N. 2006: 470	Gosepath 1992: 184
Reasonableness is relative, first of all, with regard to the respective cognitive initial state. Secondly, it is relative with regard to certain rules or standards of theoretical reasonableness.	Dadurch ergeben sich zwei Arten von Relativität der theoretischen Rationalität: zum einen eine Relativität mit Bezug auf kognitive Ausgangszustände, zum anderen eine Relativität mit Bezug auf die sog. Regeln bzw. Standards theoretischer Rationalität.

Table 3.8 Compression (g)

N. 2006: 470	Gosepath 1992: 193
A major problem of the relativistic view of reasonableness is that the expressions "reasonable" and "substantiated" lose their normal meaning. According to this view, "reasonable" is nothing more than "substantiated" for a certain person or a group of persons;	Der Relativismus ist vor allem deshalb ein Problem, weil durch ihn die Ausdrücke »rational«, »gerechtfertigt«, »begründet« ihren gewöhnlichen Sinn verlieren. Im Grunde ersetzt er sie durch Paraphrasen, die unseren Intuitionen widerstreiten. »Rational« kann gemäß relativistischer Auffasung nur noch »begründet für (eine Gruppe von) S«

References

Brill. 2019. *Brill's Publication Ethics*, July 2019. https://brill.com/fileasset/downloads_static/static_publishingbooks_publicationethics.pdf.

Dougherty, M.V. 2018. *Correcting the scholarly record for research integrity: In the aftermath of plagiarism*. Cham: Springer.

Dougherty, M.V. 2019. The pernicious effects of compression plagiarism on scholarly argumentation. *Argumentation* 33 (3): 391–412.

Fahnestock, Jeanne. 2009. *Quid Pro Nobis*: Rhetorical Stylistics for Argument Analysis. In *Examining argumentation in context*, ed. Frans H. van Eemeren, 191–220. Amsterdam/Philadelphia: John Benjamins Company.

Fox, Mark, and Jeffrey Beall. 2014. Advice for plagiarism whistleblowers. *Ethics and Behavior* 24 (5): 341–349.

Gipp, Bela. 2014. *Citation-based plagiarism detection*. Wiesbaden: Springer Vieweg.

Gosepath, Stefan. 1992. *Aufgeklärtes Eigeninteresse*. Frankfurt am Main: Suhrkamp.

Horbach, S.P.J.M., and W. Halffman. 2019. The extent and causes of academic text recycling or 'self-plagiarism'. *Research Policy* 48 (2): 492–502.

Illarietti, Davide. 2018. Il docente dell'Usi ha copiato anche il Papa. *Ticino Online*, January 18. https://www.tio.ch/ticino/attualita/1236159/il-docente-dell-usi-ha-copiato-anche-il-papa.

[N.]. 2001. Rationality as a condition for intercultural understanding. *Studies in Communication Sciences* 1: 81–99 2001.

[N.]. 2006. Comment on 'Constrained maneuvering: Rhetoric as a rational enterprise'. *Argumentation* 20 (4): 467–471. [Correction issued in *Argumentation* 2019.].

[N.] 2020. Correction to: Comment on 'Constrained maneuvering: rhetoric as a rational enterprise'. *Argumentation* 34 (2): 285.

Oransky, Ivan. 2019. Compression plagiarism: An "under-recognized variety" that software will miss. *Retraction Watch*, May 1. https://retractionwatch.com/2019/05/01/compression-plagiarism-an-under-recognized-variety-that-software-will-miss.

Palus, Shannon. 2016a. Communications researcher loses two book chapters, investigated for plagiarism. Retraction Watch, April 18. http://retractionwatch.com/2016/04/18/communicationsresearcher-loses-two-book-chapters-investigated-for-plagiarism.

Palus, Shannon. 2016b. Philosopher earns 14th retraction for plagiarism. *Retraction Watch*, June 8. http://retractionwatch.com/2016/06/08/philosopher-earns-13th-retraction-for-plagiarism.

Pemberton, Michael, et al. 2019. Text recycling. *Learned Publishing* 32 (4): 355–366.

Stern, Victoria. 2018a. A cardinal sin? Communications researcher accused of plagiarizing former Pope. *Retraction Watch*, January 12. https://retractionwatch.com/2018/01/12/cardinal-sin-communications-researcher-accused-plagiarizing-former-pope.

Stern, Victoria. 2018b. University defends researcher accused of plagiarizing former Pope. *Retraction Watch*, January 31. https://retractionwatch.com/2018/01/31/university-defends-researcheraccused-plagiarizing-former-pope.

da Silva, Teixeira, A. Jaime, and Judit Dobránszki. 2017. Compounding error. *Academic Questions* 30 (1): 65–72.

Tindale, Christopher W. 2006. Constrained maneuvering. *Argumentation* 20 (4): 447–466.

Tindale, Christopher W. 2009. Constrained maneuvering. In *Examining argumentation in context*, ed. Frans H. van Eemeren, 41–59. Amsterdam: John Benjamins Company.

van Eemeren, Frans H., and Peter Houtlosser. 2006. Preface. *Argumentation* 20 (4): 377–380.

van Eemeren, Frans H., et al. 2014. *Handbook of argumentation theory*. Dordrecht: Springer.

van Eemeren, Frans H. 2015a. Retraction note. *Argumentation* 29 (4): 493.

van Eemeren, Frans H. 2015b. Erratum. *Argumentation* 29 (4): 481–491.

Weber-Wulff, Debora. 2014. *False feathers: A perspective on academic plagiarism*. Heidelberg: Springer.

Weinberg, Justin. 2018a. Plagiarizes again—And is caught by philosophy prof.'s class (updated). *Daily Nous*, January 15. http://dailynous.com/2018/01/15.

Weinberg, Justin. 2118b. Plagiarist's university issues criticism …of the whistleblower. Daily Nous, February 1. http://dailynous.com/2018/02/01/plagiarists-universityissues-criticism-whistleblower.

Weinberg, Justin. 2019. Translation plagiarism in philosophy. *Daily Nous*, October 1. http://dailynous.com/2019/10/01/translation-plagiarism-philosophy.

Chapter 4
Dispersal Plagiarism

Abstract This chapter identifies the phenomenon of *dispersal plagiarism* and analyzes its effects. Dispersal plagiarism occurs when a lengthy single work by a genuine author is divided by a plagiarist who then publishes portions of it in various venues with insufficient credit to the genuine author. A typical case involves the appropriation of a book by a plagiarist who publishes the chapters (or sections of chapters) in discrete articles across several journals. In many instances, subsequent researchers are more likely to encounter the scholarly contributions of the genuine author in one or more of the plagiarizing versions rather than in its original book format. In the downstream literature, citations accrue to the several plagiarizing versions instead of to the original, and the body of published literature is thereby corrupted. If dispersal plagiarism is combined with duplicate publication and text recycling, as is often the case, the original author's work can virtually disappear from the research literature as citations go to the more recent and numerous plagiarizing versions. To exhibit the negative impact of dispersal plagiarism, this chapter examines three unrelated cases of the phenomenon from the published research literature in the discipline of philosophy.

Keyword Dispersal plagiarism · Duplicate publication · Text recycling · Citations · Disguised plagiarism

Dispersal plagiarism occurs when a plagiarist divides a genuine author's work and publishes portions of it in various venues without credit to the genuine author. For example, a plagiarist might take a researcher's monograph and then republish its chapters as new articles in various journals, pretending to have authored each of them and passing them off as new research. The lengthy source text is mined to create discrete plagiarizing articles. A successful performance of dispersal plagiarism results in multiple publications being credited to the plagiarist. It can quickly launch a new plagiarist into an academic career or promote a seasoned plagiarist to further undeserved academic laurels.

Once published, the plagiarizing articles begin to garner citations in the downstream literature that should be going to the original source work. Not only is the genuine author denied credit for discoveries, but the reliability of the body of published research literature is damaged. Dispersal plagiarism disrupts bibliometrics on a large scale, since researchers are more likely to encounter, and then cite, one of the many plagiarizing articles on a topic rather than the single original source work by the genuine author. The tendency of some researchers to cite the more recent relevant literature in one's field may also contribute to the original source work being eclipsed by the more recent plagiarizing articles. Within a few years, the original source work can be displaced by the various plagiarizing versions, especially if the subsequent plagiarizing articles have appeared in journals that enjoy a high visibility in an academic field.

Even if a case of dispersal plagiarism is discovered and editors and publishers issue the requisite statements of retraction for each plagiarizing article, many years may have passed between the publication of the plagiarizing versions and the discovery that they are plagiarizing. At that point, the fraudulent articles may have already become fully embedded in the body of published literature by many citations, and they may even be considered as standard works in a subfield. Furthermore, when a given article is retracted for plagiarism, it may continue to be cited by researchers who are remain unaware that a publisher has changed the status of the version of record of the article. The influence of plagiarizing articles is difficult to extirpate, and acts of dispersal plagiarism multiply on a large scale the challenges of correcting the body of published literature in the aftermath of plagiarism. In short, dispersal plagiarism is an especially pernicious form of disguised plagiarism that produces significant disorder in the body of published research literature. It also causes widespread institutional inefficiencies, since on the basis of undetected wide-scale plagiarism academic malefactors take up academic positions, grants, promotions, and awards that should have gone to authentic researchers.

Many factors complicate the process of tracking down all the plagiarizing articles in a case of dispersal plagiarism. Those who engage in dispersal plagiarism commonly add acts of duplicate publication and text recycling to their academic misdeeds. Even though republishing the plagiarizing work in additional venues increases the risk to the plagiarist that the fraud will be discovered, the multiplication of plagiarizing articles also increases the likelihood that some of the deficient articles will remain undetected and will thereby remain as part of the plagiarist's research profile. Not all institutional authorities are eager to examine the complete published corpus of a researcher who has earned a retraction for plagiarism, especially if the researcher has a lengthy list of publications. When dispersal plagiarism is combined with other forms of disguised plagiarism—such as translation plagiarism or compression plagiarism—years if not decades may pass between the publication of the plagiarizing versions and the discovery of the plagiarism. It is likely that many acts of dispersal plagiarism remain undiscovered.

Some researchers may be skeptical over whether acts of dispersal plagiarism can successfully be carried out. After all, the dispersal plagiarist must get the plagiarizing material by editors, peer reviewers, and the readership of not just one but several journals and publishers. To enjoy the professional benefits of the fraudulent publications, the dispersal plagiarist must also count on the plagiarism remaining undetected, either for the course of a career or at least for some years, until the

4 Dispersal Plagiarism

plagiarist has established enough professional clout to fend off accusations of plagiarism. The dispersal plagiarist must also deceive colleagues, as well as members of search committees, grant agencies, and institutional promotion committees responsible for reviewing and evaluating publication records of researchers. Despite these seemingly formidable hurdles, major acts of dispersal plagiarism have been carried out.

In this Chapter I analyze the phenomenon of dispersal plagiarism by examining three cases of it from the field of philosophy. I focus on three distinctive source works that were appropriated by three different plagiarists in unrelated acts of dispersal plagiarism:

- a 2003 dissertation in contemporary political philosophy.
- a 1994 monograph on late medieval and early modern ethics.
- a 1982 German-language monograph on ancient philosophy.

Collectively, these three lengthy works were mined by three separate dispersal plagiarists to produce portions of at least 17 articles and book chapters that appeared in print from 1998–2014 without credit to the genuine authors. In each case, the lengthy source work by a genuine author was fractured by a plagiarist into four or more journal articles or book chapters (or portions of journal articles and portions of book chapters). These 17 plagiarizing articles and book chapters were published in English in journals and books from established academic publishers in philosophy. To date, editors and publishers have issued corrections for 10 of the 17 (8 retractions and 2 expressions of concern) during the period of 2009–2019. I consider below how the source works were divided and passed off as original works in English-language philosophy venues and how the subsequent plagiarizing articles have been positively received in the downstream literature.

4.1 Case 1: The Clandestine Afterlife of a 2003 Dissertation

In 2003, Hans Kribbe was awarded a Ph.D. from the London School of Economics and Political Science (LSE) after completing a doctoral thesis entitled "Corporate Personality: A Political Theory of Association" (Kribbe 2003). Consisting of seven main chapters, the thesis engaged the work of John Locke, David Hume, Derek Parfit and others to defend the applicability of the notion of corporate personality to contemporary problems in ethical and political philosophy. Within just a few years, however, the thesis was mined by a professor at an English university in an act of dispersal plagiarism. The academic malefactor in this case—hereafter designated as "A."—was also a graduate of LSE and had been awarded a Ph.D. just a year after Kribbe had received his degree. Before discovery of the plagiarism in 2019, the dispersal plagiarist had succeeded in republishing much of Kribbe's thesis in several articles in well-established journals. In response to the revelation that his thesis had been appropriated, Kribbe told a journalist that he had considered that his thesis would need revision before being ready for publication, and that employment in the public sphere after graduation had prevented any work on it (Oransky 2019). Contrary to Kribbe's estimation, it turns out that the thesis was of publishable quality, since much of it was published serially in several venues without credit to Kribbe.

In this case of dispersal plagiarism, Kribbe's work was published under A.'s name in six journal articles, from 2008–2012, and all the articles consisted almost entirely of verbatim and near-verbatim repetition of portions of Kribbe's 295-page thesis, including footnotes. Nowhere in these articles did A. credit or even mention Kribbe. The articles appeared in *The Heythrop Journal* (1 article), *Ratio Juris* (2 articles), and *International Journal of Law in Context* (*IJLC*) (3 articles). Wiley, one of the largest commercial scholarly publishing houses, is the publisher of the first two journals, and Cambridge University Press, the oldest university press, is the publisher of the third journal.

In addition to committing dispersal plagiarism using Kribbe's thesis, A. also engaged in significant duplicate publication, publishing some of Kribbe's chapters more than once. Table 4.1 provides a plot-point depiction of the extent of overlap between A.'s six plagiarizing articles (A. 2008a, b, c, 2009, 2011, 2012) and the unpublished dissertation (Kribbe 2003).

Each point represents a page featuring text from Kribbe's thesis. As the table displays, this instance of dispersal plagiarism includes the publication of three chapters of Kribbe's dissertation (Chaps. 1, 3, and 6) in their entirety as separate journal articles. Furthermore, extensive duplicate publication resulted in Chap. 1 being published three times in one year (A. 2008a, b, c) and Chap. 3 being published twice in consecutive years (A. 2011, 2012). In addition to incorporating verbatim and near verbatim the entirety of Chap. 3, A. 2012 also re-produced small portions of Chaps. 2, 4, and the Conclusion. Duplicate publication in itself—independent of plagiarism—is considered a significant form of research misconduct by many publishers and researchers, since the time and resources of editors, reviewers, publishers, and readers is wasted under the pretense that new research is being offered (see Barry 2018;

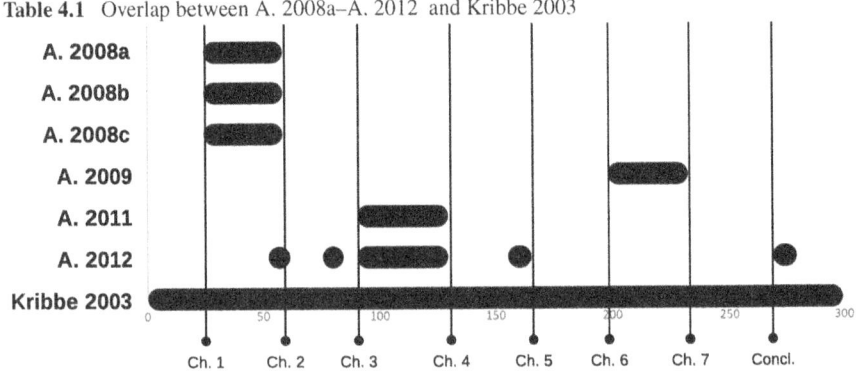

Table 4.1 Overlap between A. 2008a–A. 2012 and Kribbe 2003

4.1 Case 1: The Clandestine Afterlife of a 2003 Dissertation

Table 4.2 Summary of overlap between A. 2008a–A. 2012 and Kribbe 2003

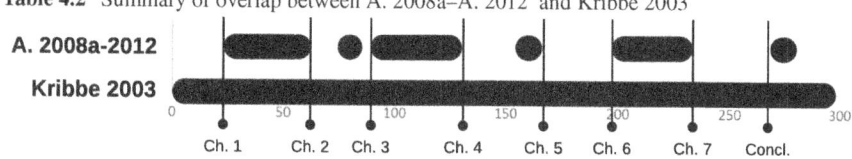

Autor 2011; Roig 2015: 16–33; Lin 2020). Even had A. 2008b, c, 2012 been unplagiarizing, two of the three would still have warranted retraction because of duplicate publication.

The magnitude of the appropriation of Kribbe's thesis in the various publications under A.'s name can be considered in several ways. Three of the seven chapters of the thesis appeared in print in their entirety, plus a few pages from two other chapters and the conclusion. If one sets aside the front matter and the bibliography of Kribbe's work, the 295-page thesis contains 277 pages of original research. Text from 90 of the 277 pages—almost a third—was published under A's name in the six plagiarizing journal articles. Table 4.2 summarizes the extent of appropriation between A. 2008a–2012 and Kribbe's thesis, presenting the known, demonstrated overlap between A.'s six articles and Kribbe's thesis.

In early 2019, the affected journal editors and publishers decided to issue retractions for the six articles in response to reporting by at least two whistleblowers. The retraction statement in *The Heythrop Journal* identified clearly that the reason for the retraction was the "large amount of plagiarised content" in the article, but the statement neglected to identify Kribbe's thesis as the hidden source text (Anonymous 2019a: 339). The three retraction statements in *IJLC* credited Kribbe's work, explaining that each article "reuses sections" of the thesis "without permission or acknowledgement" (Anonymous 2019c: 103, d: 104, e: 105). The notices in *Ratio Juris* are brief; each promotes "major overlap with previously published articles" as the reason for retraction (Anonymous 2020a: 117, b: 118).

4.1.1 The Role of Electronic Thesis Depositories

This situation of dispersal plagiarism may have been facilitated by the presence of Kribbe's doctoral thesis in the open-access online institutional repository, *LSE Theses Online*. The thesis is available for free download as a PDF with OCR formatting that allows for the text to be easily copied and pasted into another file.

In another act of plagiarism, one unrelated to the dispersal plagiarism involving Kribbe's thesis, A. took a chapter from a different doctoral dissertation present on *LSE Theses Online* and also published it in *The Heythrop Journal*. The source text for that act of plagiarism was a 1998 thesis by Simon Wigley on the topic of desert and luck in distributive justice (Wigley 1998). A's act of plagiarism republished in full, in verbatim and near-verbatim sentences with footnotes, the third chapter of

Wigley's thesis. The evidence of the plagiarism appeared online in the anonymous post-publication peer review forum PubPeer.com in early 2019. After being notified of this act of plagiarism, *The Heythrop Journal* issued a retraction statement that explicitly credited Wigley's thesis as the unacknowledged source text and explained that the article had presented a "large amount of plagiarised content" (Anonymous 2019b: 340). A's appropriation of Wigley's thesis appears to be garden-variety plagiarism rather than dispersal plagiarism, since A. does not appear to have appropriated any other chapters of Wigley's thesis in additional publications.

4.1.2 Different Titles and Competing Copyright Claimants

As noted above, A. published Chap. 1 of the Kribbe thesis in three separate journal articles in 2008. That thesis chapter was titled "The Strange Resurrection of a Long Forgotten Analogy" and it introduced the philosophical notion of corporate realism. In republishing Kribbe's chapter, A. gave different titles to each of the plagiarizing versions. Judging from the titles alone, one would not suspect that the text that follows in each of three articles is substantially identical. The titles are: "Understanding Subject(s): The Self as Corporation"; "A Suggested Basis for Legal Ontology"; and "The Illusory Nature of Ontological Status and its Implications for Legal, Moral and Social Organization." Since the three were published in 2008, the manuscript must have been under review at the respective journals at the same time or nearly the same time. The first of the three articles appeared in print in *The Heythrop Journal* in January, followed by the second in *Ratio Juris* in February, and then the last in *IJLC* in July. Given this timeframe, one must conclude that even though all three are plagiarizing articles, only the second and third articles warrant the additional designation of duplicate publication.

In addition to the problems of plagiarism and duplicate or redundant publication, the existence of the three identical plagiarizing articles raises the problem of copyright. Plagiarism and copyright violation are not the same: plagiarism is a failure of credit, and copyright violation is "a failure to obtain authorization" (Saunders 2010: 280). In the version of Kribbe's thesis on the *LSE Theses Online* website, a copyright page after the title page indicates that Kribbe is the copyright holder, stating, "Copyright in the Dissertation held by the Author" (Kribbe 2003: i). In the plagiarizing journal versions, however, copyright for the same text as Chap. 3 of the Kribbe thesis is declared for other parties. *The Heythrop Journal* and the *Ratio Juris* versions state that the copyright is held by "the author" but of course the author declared there is A., not Kribbe (A. 2008a: 423, b: 19). For the version in *IJLC*, the copyright is asserted to belong to the publisher, Cambridge University Press. The plagiarism results in three non-identical claimants to the exclusive copyright of substantially the same text: Kribbe, A., and Cambridge University Press. However, the latter two have no rights to Kribbe's thesis.

4.1.3 Instances of Plagiarism from Now-Retracted Articles

Table 4.3 presents a short selection of text from A.'s three 2008 plagiarizing articles alongside the original undisclosed source text from Kribbe's thesis. The text is from the early part of the three articles and an early part of Chap. 1 of Kribbe's thesis. The overlap with the source text is highlighted.

The text of 2008a–c is substantially identical with Chap. 1 of Kribbe's thesis. Notably, the first sentence in all four versions exhibits the first person: "I want to start by [...]." The "I" of course is the voice of Kribbe as genuine author, but when the sentence reappears in the plagiarizing version of 2008a and again in the plagiarizing duplicate versions of 2008b–c, the reader will likely assume that A. is the one who is speaking. Some minor changes can be found in the plagiarizing versions. In Kribbe's first sentence, the expression "to set out" has been modified as "to paint" in 2008c. Additionally, the word "relatedness" has been changed to "relations" in 2008c.

There is one minor yet revealing modification to Kribbe's text as found in 2008a and 2008c. In the original passage, Kribbe is describing adherents to a school of thought on personal identity inspired by Locke and Hume, and then Kribbe identifies some "modern members" of that school. In 2008a and 2008c, however, the corresponding passage does not have "modern members" but the unintelligible expression "modem members." Why is this? Those familiar with electronic texts will recognize that a common error made by optical character recognition programs is the failure to identify accidental ligatures; sometimes the adjacent letters "rn" are rendered as the single letter "m" when the OCR program produces an electronic version of a text from a scanned document. It this case, an OCR program appears to have made this error, generating the word "modem" instead of "modern." Indeed, in the OCR version of Kribbe's thesis on the *LSE Theses Online* website, the word "modern" appears 10 times, and in 8 of those cases, including the passage cited here, the OCR version has registered the word errantly as "modem". Since "modem" is itself an

Table 4.3 Triplicate publication of a source text

A. 2008a: 424	A. 2008b: 20	A. 2008c: 64	Kribbe 2003: 31
In order to set out the backdrop to the argument, I want to start by drawing a broad distinction between two schools of thought in the philosophy of personal identity. The first school is inspired by the ideas of Locke and Hume, while its modem members include Grice, Parfit, Perry and Quinton. It advances what is often described as the psychological criterion of personal identity. According to this criterion, personal identity over time consists in the relatedness between a series of mental states at different moments.	In order to paint the backdrop to the argument, I want to start by drawing a broad distinction between two schools of thought in the philosophy of personal identity. The first school is inspired by the ideas of Locke and Hume, while its modern members include Grice, Parfit, Perry, and Quinton. It advances what is often described as the psychological criterion of personal identity. According to this criterion, personal identity over time consists in the relatedness between a series of mental states at different moments.	In order to set out the backdrop to the argument, I want to start by drawing a broad distinction between two schools of thought in the philosophy of personal identity. The first school is inspired by the ideas of Locke and Hume, while its modem members include Grice, Parfit, Perry and Quinton. It advances what is often described as the psychological criterion of personal identity. According to this criterion, personal identity over time consists in the relations between a series of mental states at different moments.	In order to set out the backdrop to the argument, I want to start by drawing a broad distinction between two schools of thought in the philosophy of personal identity. The first school is inspired by the ideas of Locke and Hume, while its modern members include Grice, Parfit, Perry and Quinton. It advances what is often described as the psychological criterion of personal identity. According to this criterion, personal identity over time consists in the relatedness between a series of mental states at different moments.

English word, it would not have been flagged as errant by a spell-checking program. This idiosyncratic error was transferred seamlessly into the plagiarizing versions of the 2008a and 2008c from Kribbe's chapter, producing the unintelligible expression "modem members." This error was apparently neither caught by A. nor subsequently by the article production teams for Wiley and Cambridge University Pres, the two publishers of the journal articles.

Two of the "modem" for "modern" instances in the OCR version of the Kribbe thesis occur in Chap. 6. When Kribbe published that chapter in *IJLC* the OCR error is also preserved in the published journal version, so that the unintelligible phrase "modem mechanism" twice appears instead of Kribbe's "modern mechanism" (A 2009: 101, n. 22; Kribbe 2003: 219, n. 24). Still further, the error is found twice in A.'s plagiarizing version of Simon Wigley's dissertation in *The Heythrop Journal* ("modem expression" A. 2010: 659; Wigley 1998: 73; "modem predicament" A. 2010: 669, n. 4; Wigley 1998: 74, n. 41). These findings suggest that A. was copying and pasting from the electronic versions of the theses in manufacturing the plagiarizing manuscripts for submission. Again, how these idiosyncratic errors persisted through the copyediting stages at the publishers remains a mystery.

As noted above, A. engaged in plagiarism and duplicate publication a second time by publishing Chap. 3 of Kribbe's dissertation in a pair of articles that appeared in 2011 and 2012. The titles of those two plagiarizing articles are similar ("Corporate Personality: A Politico-Jurisprudential Argument"; "An Intentional Basis for Corporate Personality"), and both bear a resemblance to the title of Chap. 3 of Kribbe's thesis, "Corporate Personality: A Political Theory." Table 4.4 presents a selection from the conclusion of those articles and the conclusion of Kribbe's chapter.

Table 4.4 Duplicate publication of a source text

A. 2011: 491–492	A. 2012: 429	Kribbe 2003: 124–125
Conclusion	Conclusion	Conclusion
It can be argued that the identity of persons over time should be understood entirely in terms of a bundling of mind-states without reference to a physical or spiritual substratum, that this bundling should be in part defined in normative terms of integrity. This conception of persons is itself irreducibly practical (see [A.] 2009). If such an argument is correct, then we face the following two challenges to get the idea of a *corporate* person up and running. First, we have to present an understanding of human association that corresponds to the idea of integrity. Second, because the model depends on its normative impact, we also need to show that this understanding of human association has practical benefits. In this article, I have tackled the first challenge. I intended to show that the notion of integrity could be applied, not just to the temporal stages of a life, but also to the relatedness between individuals.	It can be argued that the identity of persons over time should be understood entirely in terms of a bundling of mind-states without reference to a physical or spiritual substratum, that this bundling should be in part defined in normative terms of integrity. This conception of personal is itself irreducibly practical.[50] {[A.] (2009, pp. 93–106).} If such an argument is correct, then we face the following two challenges to get the idea of a *corporate* person up and running. First, we have to present an understanding of human association that corresponds to the idea of integrity. Second, because the model depends on its normative impact, we also need to show that this understanding of human association has practical benefits. In this article, I have tackled the first challenge. I wanted to show that the notion of integrity could be applied, not just to the temporal stages of a life, but also to the relatedness between individuals.	I argued in Chapter I and II that the identity of persons over time should be understood entirely in terms of a bundling of mind-states without reference to a physical or spiritual substratum, that this bundling should be in part defined in normative terms of integrity, and that this conception of personal integrity is itself irreducibly practical. If these arguments are correct, then we face the following two challenges to get the idea of a *corporate* person up and running. First, we have to present an understanding of human association that corresponds to the idea of integrity. Second, because the model depends on its normative impact, we also need to show that this understanding of human association has practical benefits. In this chapter, I tackled the first challenge. I wanted to show that the notion of integrity could be applied, not just to the temporal stages of a life, but also to the relatedness between individuals.

4.1 Case 1: The Clandestine Afterlife of a 2003 Dissertation   59

The passages are nearly identical, but there are some small changes. These changes reflect the modification of the source text from the context of a dissertation to the context of a published article that purports to present new research. The original expression "In this chapter […]" has been modified in the plagiarizing versions to "In this article […]." To have kept Kribbe's exact rendering would have disclosed to readers (and reviewers and editors) that the text was not original and had already appeared in the thesis. Furthermore, the expression "I argued in Chaps. I and II […]" has been rendered as "It can be argued […]." Again this modification appears necessary to conceal the origin of the text. The word "arguments" has been reduced to the singular, because only one of the chapters is being appropriated here, and the change was necessary to avoid an implicit reference to the other chapter of Kribbe's work. No change, however, has been made to Kribbe's use of the first person, as the "I" of Kribbe is presented as the "I" of "A." in the plagiarizing articles.

In copying from the thesis, A. appears to have accidentally omitted a word, since Kribbe's expression "conception of personal integrity" is rendered unintelligibly without a substantive as "conception of personal" in A. 2012. In the corresponding passage in A. 2011, the phrase has been modified to "conception of persons." Table 4.4 also displays that there is a new citation that shows up in both plagiarizing versions that is not found in the source text: A. provides a reference to the previously published article, A. 2009. That work is simply the now-retracted article in which A. had republished Chap. 6 of the Kribbe thesis in *IJLC*. The self-citation to a plagiarizing article is notable. Why would one want to draw attention to a plagiarizing work? Some research integrity theorists have speculated that one "who plagiarizes a published article does not want to show off the article to the world as if it was his or her own. He or she wants to show off a longer publication list to the dean" (Biagioli 2012: 463). For a plagiarist to direct further attention to plagiarizing articles is risky, as doing so can increase the likelihood that the plagiarism will be identified by a reader who is familiar with the source text.

Since the source text is Kribbe's 2003 thesis, the reference list consists almost entirely—with two exceptions—of works published before 2003, a decade before the appearance of A.'s plagiarizing 2012 article. The added reference to A's plagiarizing 2009 article supports the illusion that the literature review is up to date with references to new scholarly research. Figure 4.1 demonstrates that the bulk of the literature cited in A. 2011, 2012 was published prior to 2003, the year of Kribbe's thesis.

Of the 34 entries to the work of others in the bibliographies of A 2011, 2012, all but 32 are prior to 2003. Disproportionate dependency on the older research literature can be an important contributing piece of evidence in evaluating the strength of a given plagiarism claim.

One portion of A. 2011 does not appear anywhere in Kribbe's thesis. At bottom of the first page is a note that states "Earlier versions of this article were delivered as papers at St. Edmund's College, Cambridge and the University of Frankfurt." (A. 2011: 471) The note also says, "The author wishes to thank colleagues at the Max-Planck-Institut für ausländisches und internationales Strafrecht in Freiburg for their input into this article" (ibid.). At face value these acknowledgments suggest

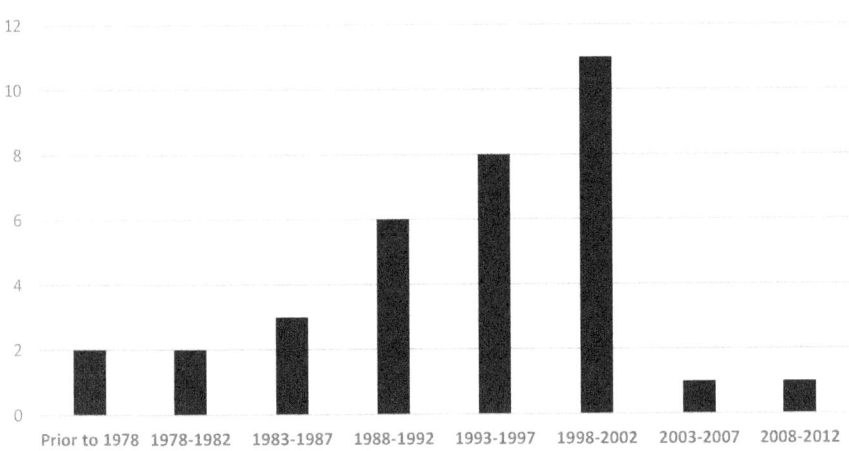

Fig. 4.1 Distribution of bibliographical references in A. 2011, 2012

that Kribbe's chapter was not only presented publicly as lectures or conference presentations but that the text was circulated for comments. If any significant comments or suggestions were rendered, they do not appear to have been incorporated into the text, since it is substantially the same as third chapter of Kribbe's (2003) thesis.

Table 4.5 presents a selection of text from A. 2009 that appropriated Chap. 6 of Kribbe's dissertation. Since Kribbe apparently only published this chapter as one journal article, the additional infraction of duplicate or redundant publication does not obtain here.

The source text is a bit lengthier than the plagiarizing version, as A. has excised Kribbe's references to other chapters of the dissertation. A.'s use of the first person and

Table 4.5 Plagiarism without duplicate publication

A. 2009: 105–106	Kribbe 2003: 227–228
We have also seen that there is a sense in which ethical individualism could have been right. Our agency could have been fractured alongside our corporeal existence. If so, every individual human being would have been a separate agent, and this is often believed to be the case. But there is also a sense in which this view need not be right or at least not exclusively right. Our agency could still be united by the normativity of collective intention. Corporate personality is a logical possibility. This is a normative question, a question that must be tested against our intuitions and the goings-on that are constitutive of our world and lives. There may be a logically possible world in which moral individualism is exclusively right, but in this paper, I argue that this is not the case in this world. My argument is that we do not merely care about ourselves, or even about our contemporaries, but look back and forward in time, much further than our own lives. We are immersed not only in our own life-plans, but also in plans and projects with a greater history and continuity. Associativism offers a genuine alternative. Corporate personality allows us to compare distributions between generations to distributions between the stages of one person.	We have also seen that there is a sense in which ethical individualism could have been right. Our agency could have been fractured alongside our corporeal existence. If so, every individual human being would have been a separate agent, and this is often believed to be the case. But there is also a sense in which this view need not be right or at least not exclusively right. Our agency could still be united by the normativity of collective intention. As I argued in Chapter III, corporate personality is a logical possibility. Which view is true? Again, this is a normative question, a question that must be tested against our intuitions and the goings on that are constitutive of our world and lives. There may be a logically possible world in which moral individualism is exclusively right, but in Chapter V, VI and VII, I argue that this is not the case in this world. A basic premise of my argument in these chapters is that we do not merely care about ourselves, or even about our contemporaries, but look back and forward in time, much further than our own lives. We are immersed not only in our own life-plans, but also in plans and projects with a greater history and continuity. In Chapter V, I argue that ethical individualism cannot account for this connectedness with the future. In this chapter, I argue that associativism offers a genuine alternative. Corporate personality allows us to compare distributions between generations to distributions between the stages of one person.

the repetition of the expression "my argument" is noteworthy, as the argumentation is indisputably identical to Kribbe's thesis. The substitution of "this paper" for "chapter" is necessary to sustain the illusion of original work and to avoid an implicit reference to the undisclosed source text.

4.1.4 The 2003 Kribbe Thesis in the Body of Published Research Literature

Since the source text of A.'s six plagiarizing articles—Kribbe's doctoral dissertation—was an unpublished thesis, it was not in a full sense part of the body of published research literature. Dissertations can be liminal works, in-between the so-called "grey" literature and the published scholarly literature, although in some countries publication is necessary for the conferral of the degree. Kribbe's thesis did not have citations in the scholarly research literature, although now it is cited explicitly in several of the retractions that were published for A.'s articles. Of course, as exemplified in the plot-point depictions in Tables 4.1, 4.2, Kribbe's thesis has enjoyed a widespread presence in the body of published research literature under A.'s name in the six plagiarizing articles.

Kribbe's work is also covertly present in part in the body of published research literature as cited and extracted by authors who engage A.'s work. Discussion of A.'s six articles in the downstream literature are really discussions of Kribbe's thesis. In addition to self-citations by A., the six plagiarizing articles have been cited by many researchers. To take one instance: in a 2019 monograph on the legal personhood of animals, Christopher Hutton observes that

> [A.] likewise argues that "The first claim is that individuals have the same ontological status as associations", finding a parallel between the individual and the collective: "Collectives, of course, are not a free-floating consciousness. They should be broken down into individually conscious parts. But on the other hand, a more thoroughgoing ontological reductionism uncovers that individuals are in turn like miniature associations" (2008: 63). [Hutton 2019: 98]

The two sentences that Hutton here attributes to A. are from the first chapter of Kribbe's thesis (Kribbe 2003: 29–30). Hutton is unknowingly citing Kribbe's work through the proxy of A. 2008c—the article in *IJLC*—which was the second duplicate publication (or "triplicate publication") of A.'s previous plagiarism of Chap. 1 of Kribbe's thesis (after appearing in A. 2008a and then in A. 2008b). Figure 4.2 demonstrates the progression of Kribbe's words from their origin in the 2003 thesis to their citation by Hutton sixteen years later.

The words of Kribbe (1) are present in the literature as initially plagiarized (2), as redundant in duplicate (3) and triplicate publication (4), and finally as cited in the downstream literature (5). Hutton's readers are likely unaware how well-travelled the words are prior appearing in Hutton's 2019 monograph.

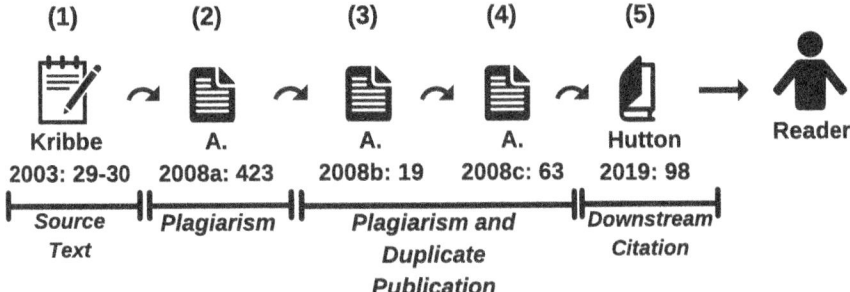

Fig. 4.2 The journey of two sentences from Kribbe 2003

A.'s misuse of Kribbe's thesis is a clear, instructive instance that illustrates the phenomenon of dispersal plagiarism and the downstream negative effects of the phenomenon. As far as such cases go, it was resolved reasonably well, since each of affected publishers issued retractions for the six plagiarizing articles. When publishers issue retractions for only a subset of the plagiarizing articles in a case of dispersal plagiarism, or none at all, however, the harm to the body of published research literature can endure.

4.2 Case 2: The Post-publication Re-Emergence of a 1994 Monograph

In 1994, Ilkka Kantola published his book *Probability and Moral Uncertainty in Late Medieval and Early Modern Times* with the small Finnish academic publishing house Luther-Agricola-Society. Consisting of 205 pages divided into 4 main chapters, the study considers the history of the moral theory of probabilism as expressed by philosophers and theologians of the stated period. Kantola did not pursue an academic career but entered public service, first as a Lutheran minister and bishop, and then as an MP in the Finnish parliament. Only in 2009 did he become aware, after being informed by some researchers, that portions of his book had been appearing in journal articles and book chapters under the name of a seemingly productive European philosophy professor (hereafter, "S.").

In this instance of dispersal plagiarism, seven journal articles and book chapters by S. re-presented material from Kantola's 1994 book, and they appeared in print from 1999–2005. As it would turn out, these seven were part of a larger serial plagiarism situation involving 45 plagiarizing articles and chapters by S. published during a 14-year period (see Dougherty, Harsting, and Friedman 2009; Dougherty 2017). The focus here, however, is on a subset of these 45 articles, namely, the seven plagiarizing works that together constitute a case of dispersal plagiarism by S. of Kantola's monograph.

Table 4.6 presents a plot-point depiction of the seven plagiarizing articles and

4.2 Case 2: The Post-publication Re-Emergence of a 1994 Monograph

Table 4.6 Overlap between S. et al. 1999–S. 2005 and Kantola 1994

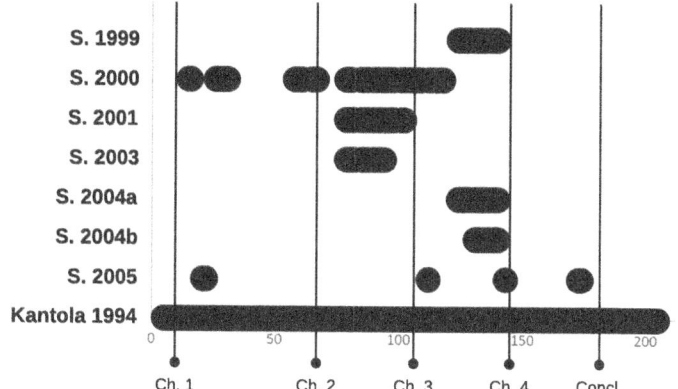

Table 4.7 Summary of overlap between S. et al. 1999–S. 2005 and Kantola 1994

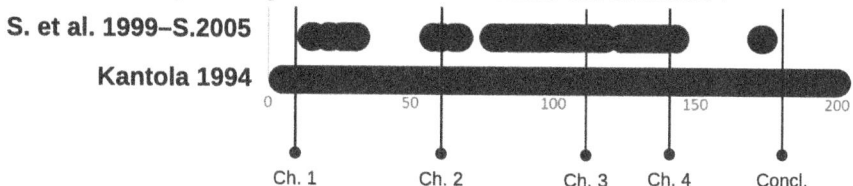

book chapters by S. et al. (1999), S. (2000), (2001), (2003), (2004a, b), (2005) and Kantola's monograph (Kantola 1994).[1]

The extent of dependency on Kantola's book varies in the seven plagiarizing works. S. 2000, a journal article, consists almost entirely of undocumented extracts (including footnotes) from Kantola's book, whereas S. 2005, a book chapter, appropriates only a few very short selections. Again, the phenomenon of dispersal plagiarism coincides with duplicate publication, or at least, substantial text recycling: significant portions of Chap. 2 of Kantola's book reappear in three plagiarizing works by S. (2000), (2001), (2003) and significant portions of Chap. 3 reappear in three other plagiarizing works (S. et al. 1999; S. 2004a, b).

Although no individual chapter of Kantola's book was published in its entirety as a discrete article by S., a significant total portion of the book eventually appeared in print under S.'s name across the seven publications. Not counting the front and back matter of Kantola's 205-page book, there are 181 pages of original research set forth in the four chapters and the conclusion. Text from 130 of those 181 pages appear in the seven journal articles and book chapters by S. Table 4.7 presents a summary of the overlap between S et al. 1999–S. 2005.

---

[1] The data for this table is extracted from Dougherty, Harsting, and Friedman (2009): 356–357, 362, 368, 371–372, 376.

Journal editors and publishers responded in varied ways to the disclosure of the plagiarism. The two earliest plagiarizing works (S. et al. 1999; S. 2000) were journal articles, and retractions appeared soon after the plagiarism was reported. The retraction statement for the first explained that "serious plagiarism had been detected" and that the "plagiarism represents a serious assault on intellectual integrity" (ETL 2010: iii). It also specified that although the article was co-authored, the "plagiarism in question was related exclusively to the portion of the article for which [S.] was responsible" (ibid.). The second retraction statement declared that the "evidence of plagiarism" was "overwhelming and irrefutable" (RTPM 2009: v). For the remaining five plagiarizing works by S. (all book chapters) two publishers issued what could be described as neutral expressions of concern (Salles 2010: ix, 517, 543; Anonymous 2010), and the three remaining publishers issued no statements.

4.2.1 Examples of Plagiarism

Table 4.8 presents a passage found in three articles by S. alongside the original source text from Chap. 2 of Kantola's book. The subject matter concerns the philosopher Francisco Suárez's views on how an agent should act when faced with a decision between competing alternatives that each are endorsed by reputable authorities.

After the initial act of plagiarizing the passage in S. et al. (1999) the plagiarizing text was recycled again in two more publications (S. 2004a, b). Some minor modifications have been made to Kantola's source text. Many of the changes are mere synonym substitutions. These changes do not affect the substance of Kantola's analysis of Suárez's view probabilistic reasoning.

A similar level of minor modifications can be found in S.'s appropriation of Chap. 2, where Kantola presents an account of practical reasoning as found in works by several medieval philosophers and theologians. Table 4.9 presents a selection of Kantola's discussion of Henry of Ghent's view alongside three plagiarizing versions (S. 2000, 2001, 2003).

Again, there are minor English synonym substitutions in the three plagiarizing versions, but the main difference is the simple replacement throughout of the Latin expressions "*recta ratio*" and "*conscientia*" for "right reason" and "conscience."

4.2.2 Kantola Versus S. in the Downstream Literature

The content of Kantola's book has a twofold existence in the body of published literature: in the 1994 book itself and in S.'s seven plagiarizing articles. (In contrast, the Kribbe thesis is represented in the scholarly literature under Kribbe's name only as mentioned in retraction statements for A.'s plagiarizing articles.) One can compare how the content of Kantola's research has fared in the downstream literature in the two distinct ways in which it has been disseminated: as explicitly published under

4.2 Case 2: The Post-publication Re-Emergence of a 1994 Monograph

Table 4.8 Kantola's account of Suárez in three works in triplicate publication by S

S. 1999 et al.: 379	S. 2004a: 13–14	S. 2004b: 53	Kantola 1994: 137–138
Suarez then comes to that problematic situation in which one must make a decision between two probable opinions, of which one is more probable than the other? At first, Suarez makes a distinction between the probable opinions of law ("opiniones circa ius") and the probable opinions of fact ("opiniones circa res"). When there are two opposing probable opinions in regard to the existence of a command of law, and one opinion supports the view that such a command does not exist, one is, according to Suarez, able and permitted to form a sure practical judgement of conscience, basing it on this probable opinion. One is allowed to assume the risk of acting against the law. It would be unfair to require that everyone should make comparisons between different probable opinions in order to decide which one of them is the best candidate for action. All one needs to know in order to act is that opinions which will ground the action are genuinely probable opinions. Suarez's view, then, is that one can proceed in one's practical reasoning in the following manner: I know that there is a probable opinion to the effect that this kind of act is not prohibited; therefore, I am "sure" that I am allowed to perform that act.	Given Suarez's position what, then, is his solution to that problematic situation in which one must make a decision between two probable opinions, one of which is more probable than the other? At first, he makes a distinction between the probable opinions of law (*opiniones circa jus*) and the probable opinions of fact (*opiniones circa res ipsas*). When there are two opposing probable opinions in regard to the existence of a command of law, and one opinion supports the view that such a command does not exist, an agent is permitted to form a sure practical judgement of conscience, basing it on this probable opinion. Suarez holds that an agent is allowed to assume the risk of acting against the law, since it would be unfair to require that everyone should make comparisons between different probable opinions in order to be able to decide which of them is the best candidate for action. All one needs to know in order to act is which opinions are probable and which are not. Suarez's view is that one can proceed in one's practical reasoning in the following way: I know that there is a probable opinion to the effect that this kind of act is not prohibited; therefore I am 'sure' that I am allowed to perform that act.	Given Suarez's position what, then, is his solution to that problematic situation in which one must make a decision between two probable opinions, one of which is more probable than the other? His answer is grounded in a distinction between the probable opinions of law (*opiniones circa ius*) and the probable opinions of fact (*opiniones circa res ipsas*). When there are two opposing probable opinions in regard to the existence of a command of law, and one opinion supports the view that such a command does not exist, an agent is able and is permitted to form a sure practical judgement of conscience, basing it on this probable opinion. Suarez holds that one is allowed to assume the risk of acting against the law, since it would be unfair to require that everyone should make comparisons between different probable opinions in order to be able to decide which of them is the best candidate for action. All one needs to know in order to act is which opinions are probable and which are not. Suarez's view is that one can proceed in one's practical reasoning in the following way: I know that there is a probable opinion, to the effect that this kind of act is not prohibited; therefore, I am (practically) 'certain' that I am allowed to perform that act.	What is, then, Suárez's solution to the problematic situation in which one must make a decision among two opposing probable opinions, of which the one is more probable than the other? [...] At first, Suárez makes a distinction between probable *opinions of law* (*opiniones circa ius*) and the probable *opinions of fact* (*opiniones circa res*). When there are two opposing probable opinions in regard to the existence of a command of law, and one opinion supports the view that such a command does not exist, one is, according to Suárez, able — and allowed — to form a sure practical judgment of conscience, basing it on this probable opinion. Suárez appears to believe that one is allowed to assume the risk of acting against the law. He states that it would be unfair to require that everyone should make comparisons between different probable opinions in order to be able to decide which one of them is the most probable one. It is sufficient that one knows which opinions are probable and which are not. Suárez's view is that one can proceed in one's practical reasoning in the following way, "I know that there is a probable opinion to the effect that this kind of action is not prohibited, therefore I am certain that I am allowed to perform that action."

Kantola's name and as fraudulently published under S.'s name. *Google Scholar* presently lists fewer citations to Kantola's book from 1994–2019 than to the seven plagiarizing articles by S. from 1999–2019. One can draw the unfortunate conclusion that Kantola's research has been slightly more influential in the plagiarizing versions by S. than in its authentic monograph format.

Some oddities have arisen in the way researchers have referenced Kantola's book and the plagiarizing versions of it by S. Some researchers have cited both Kantola and S.'s plagiarizing works together, sometimes within a single footnote, apparently not realizing the identity of the two (e.g., Maryks 2008: 65, n. 76; Schüßler 2006: 160, n. 33; Schüssler 2005: 92, n. 2; see Astorri 2019). Furthermore, the two journal articles that plagiarize Kantola's work (S. et al. 1999; S. 2000) have continued to receive citations even after they were retracted. This fact is especially surprising given the

Table 4.9 Kantola's account of Henry of Ghent in triplicate publication by S

S. 2000: 131	S. 2001: 807	S. 2003: 214	Kantola 1994: 91
The judgement of the practical intellect represents *recta ratio* (right reason) in regard to particular action; it is formed through applying universal moral principles to a particular situation. Thus *recta ratio* concerns itself with particular moral knowledge. Henry states, however, that having such moral knowledge does not necessarily imply having a conscience in regard to particular action. One may have moral knowledge of a particular action without necessarily having a conscience in relation to that action. This means that an agent in possession of an erroneous moral reason does not necessarily possess an erroneous conscience. *Recta ratio* and *conscientia* (conscience) differ in the sense that while recta ratio pertains to the cognitive part of the soul, *conscientia* pertains to the affective part of the soul.	The judgement of the practical intellect represents *recta ratio* (right reason) in regard to a particular action; it is formed by applying universal moral principles to a particular situation. Thus, *recta ratio* concerns itself with detailed moral knowledge. The possession of such moral knowledge, for Henry, does not necessarily imply having a conscience in regard to particular action. One may have moral knowledge of a particular action without necessarily having a conscience in relation to that action. This entails that an agent in possession of an erroneous moral reason does not necessarily possess an erroneous conscience. *Recta ratio* and *conscientia* (conscience) differ in the sense that while *recta ratio* pertains to the cognitive part of the soul, *conscientia* pertains to the affective part.	The judgement of the practical intellect represents *recta ratio* (right reason) in regard to particular action, and is formed through applying universal moral principles to a particular situation. Thus *recta ratio* concerns itself with particular moral knowledge. Henry states, however, that having such moral knowledge does not necessarily imply having a conscience in regard to particular action. One may have moral knowledge of a particular action without necessarily having a conscience in relation to that action. This means that an agent in possession of an erroneous moral reason does not necessarily possess an erroneous conscience. *Recta ratio* and *conscientia* (conscience) differ in the sense that while *recta ratio* pertains to the cognitive part of the soul, *conscientia* pertains to the practical part of the soul.	The judgement of practical intellect represents the *right reason* in regard to particular action; it is formed through applying universal moral principles within a particular situation. Thus *right reason* means particular moral knowledge. Henry, however, states that having this particular moral knowledge does not necessarily imply having a conscience in regard to a particular action. One may have moral knowledge of a particular action without necessarily having a conscience in relation to that action. According to Henry, this means that neither does having an *erroneous moral reason* imply having an *erroneous conscience*. Right reason and conscience differ in the sense that while the right reason pertains to the cognitive part of the soul, conscience is relevant to the affective part of the soul.

lengths to which the publisher of both has gone to correct the scholarly record. Not only were the clear retractions issued for both articles, but the electronic versions were removed publisher's website. Additionally, each retraction notice describes that self-adhesive labels indicating plagiarism were being sent to all journal subscribers so that the labels could be attached to the first and last pages of the articles to warn future readers (see RTPM 2009: v; ETL 2010: iii). These extensive actions, however, have not wholly succeeded in stemming citations to the two now-retracted articles. They continue to accrue citations (e.g., Hallebeek 2018: 232, n. 31; Holland 2017: 51–52, n. 83).

S. (2000) recently received an odd citation in a journal article that includes an extensive 9-line extract from the retracted article (Pine 2019: 724, n. 26). The author does not indicate that S. (2000) has been retracted, but appends a note after the quotation that states, "*Nota bene*: [S.] has been shown to be guilty of serial plagiarism; I cite [S.'s] work with this caveat" (ibid.). Referencing S. (2000) for reasons other than to discuss plagiarism may be unnecessary, since researchers can instead cite the source articles, which have been extensively documented for all of S.'s plagiarizing works (Dougherty, Harsting, and Friedman 2009). Nevertheless, continued positive citations to S. (2000) continue (e.g., Korzo 2019: 65, n. 10, 69, n. 22).

Perhaps some continuing citations to S. are unavoidable. One recent instance involved the re-publication in a new language of a deceased scholar's older work that originally cited one of the now-retracted articles (see Vismara 2018: 617, n. 20).

Furthermore, S. (2000) is present in the JSTOR repository with no evidence that it has been retracted and removed from the publisher's website (JSTOR 2000). When retraction statements appear with articles only on a subset of venues through which researchers access scholarly papers, ongoing citations to faulty research continue. For more on this problem in general, see Dougherty 2018: 226–227.

The increased role of bibliometrics in some scientific fields has led some researchers to conclude that articles retracted for plagiarism and other acts of research misconduct should never be cited directly. Otherwise, a benefit is conferred to academic malefactors. Nevertheless, discussions of research misconduct require that that such articles be cited in discussion of scientific integrity (Thielen 2018: 188). One author who considered some of S.'s retracted articles does so obliquely, noting that "retracted articles mentioned in this article are not included in the reference list. This is done in order to avoid their continuous citations" (Halevi 2020: 61; Fig. 3, p. 60 in that article refers to one of S.'s retracted articles).

4.3 Case 3: The English Re-Incarnation of a German Monograph

The two preceding cases of dispersal plagiarism featured English-language source texts that were mined by plagiarists to produce plagiarizing book chapters and articles. A more subtle form of dispersal plagiarism occurs when it is combined with translation plagiarism. The change in language for the plagiarizing versions creates a greater distance between them and the source text. (See Chap. 2 above.) This distance undoubtedly lessens the prospect of discovery, since detection is restricted to researchers capable of doing work in both languages. As noted previously, demonstrating instances of translation plagiarism can be particularly challenging, since an exact word-to-word correspondence may not be preserved as the source text moves from its source language to one with idiomatic ways of expressing concepts and different syntactical structures. The last case of dispersal plagiarism to be considered here involves the dissemination of a German-language monograph in plagiarizing English-language journal articles and book chapters.

In 1982 philosopher Wolfgang Wieland published his lengthy study of Plato, *Platon und die Formem des Wissens*. Early reviewers raised critical objections to the book, yet still recognized it as a substantial contribution to the understanding of Plato's thought (e.g., Taylor 1983; Ryan 1984; Frede 1986). The success of the 339-page book led to its republication in 1999, with no change to the original body of the text with the exception of two short new essays that were appended to the end of the book. A prolific researcher on ancient Greek philosophy, Wieland published influential work on Aristotle in English, but his study on Plato has remained untranslated. At least, the monograph has not appeared in English under his own name.

Beginning in 2002, a European philosophy professor—hereafter, "N."—began to publish portions of Wieland's 1982 book in English translation in the guise of

Table 4.10 Overlap between N. 2002–N. 2014 and Wieland 1982

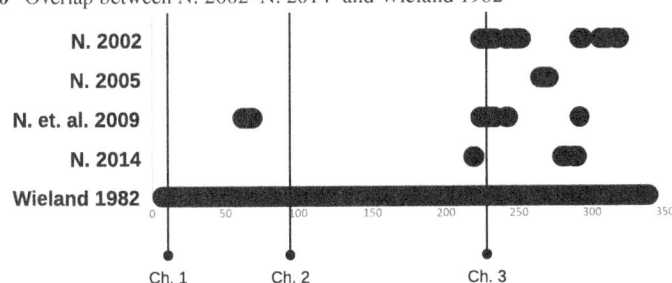

new original scholarship.[2] This case of dispersal plagiarism involved at least four plagiarizing publications: two journal articles and two book chapters (N. 2002, 2005, 2014; N. et al. 2009). The dependency of each of these works on Wieland's monograph varies significantly. Nearly the entirety of N. 2002 is an unacknowledged direct translation of passages from Wieland's book masquerading as an original article of scholarship. Seven pages of a co-authored article—N. et al. 2009—originate from Wieland's book. Only short sections of Wieland's book appear in N. 2005, 2014. There are no references to Wieland's book in the first three plagiarizing works. In the last, it is referenced, but only as an entry in the bibliography on the last page. This case of dispersal plagiarism extends over considerable period, since more than three decades passed between the original publication of the German source text and the most recent plagiarizing re-appearance of portions of it in N. 2014. Of the four plagiarizing works, none has been retracted. N. 2005 has received an online corrigendum from the publisher that admits the chapter "has certain severe shortcomings in the references" yet oddly asserts the view that N. "never copied the entirety or essential parts of other people's publications" (Anonymous 2017). No source texts are identified in this statement. A January 2016 posting about N. 2005 on PubPeer.com identifies five unacknowledged source works, but they are each English-language sources.

In this situation of dispersal plagiarism, the publication of some of the plagiarizing material constitutes instances of text recycling, since some portions of the source text show up in more than one of the plagiarizing articles. Table 4.10 reveals via a plot-point depiction the extent of overlap between the four plagiarizing articles by N. and Wieland 1982, where each point represents a page containing text by Wieland's book.

The portions that N. 2002, and N. et al. 2009 have in common are taken from the last portion of Chap. 2 of Wieland's book, the early portion of Chap. 3, and a short portion from the middle of Chap. 3. As can be seen, the bulk of the text of Wieland's book that is incorporated without credit in the works of N. is taken

---

[2] I am grateful to Pernille Harsting for sharing with me her discovery that Wolfgang Wieland's 1992 monograph, *Platon und die Formen des Wissens*, is the source text for nearly all of N. 2002 and much of N. et al. 2009. Her discoveries led me to discern that N. 2005 and N. 2014 borrow small portions from Wieland's book.

4.3 Case 3: The English Re-Incarnation of a German Monograph

Table 4.11 Summary of overlap between N. 2002–N. 2014 and Wieland 1982

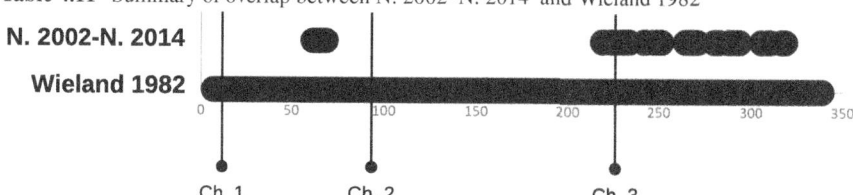

from the second half of the 1982 monograph, especially from Chap. 3, which is the concluding chapter on the forms of knowledge. Much of Wieland's discussion there is not a direct exposition of Plato's text, but a general epistemological account inspired by insights from contemporary philosophers. For this reason, apparently, it was able to be fashioned by N. to form parts of articles that are not primarily concerned with the thought of Plato. This feature, in addition to the concealing factor of the change from German to English, may partially explain why the plagiarism remained undiscovered for so long.

Table 4.11 provides a summary of the overlap between N. 2002–N. 2014 and Wieland 1982. Apart from the front and back matter of Wieland's lengthy 339-page book, there are 318 pages of original research. Text from 55 of those pages appear in English translation in N. 2002–N. 2014.

4.3.1 Dispersal Plagiarism in Translation

Judging only from the titles of N.'s four plagiarizing works, one might be skeptical that a generally well-regarded book on Plato could be the source text for passages in plagiarizing works that themselves are not mainly about Plato. Of the four, N. 2002 and N. et al. 2009 appropriate the most text from Wieland's book. N. 2002 was published in *Semiotics*, the annual yearbook of the Semiotic Society of America, and is titled "'Know How' and 'Know That': A Semiotic Approach to the Phenomenon of Knowledge." N. et al. 2009 was published in *Semiotica*, the journal of the International Association for Semiotic Studies, and is titled "Subjectivity out of Irony." Table 4.12 presents a selection from Chap. 3 of Wieland's book with the corresponding text as found in the plagiarizing articles. The overlap of the English rendering of the uncited German source text is indicated by highlighting. Despite the translation to another language, the passages are substantially the same with only very minor additions in the English.

The subject matter of the source passage is not exclusively concerned with Plato's views, and hence it was able to be conscripted into plagiarizing service for the two semiotics articles. That much of Chap. 3 of Wieland's book is about general epistemological matters rather than Plato's views in particular was noted by early reviewers, including one who characterized it as where Wieland "tries to give the necessary

Table 4.12 Translation plagiarism and text recycling on propositional statements

N. 2002: 378	N. et al. 2009: 407–408	Wieland 1982: 225
The methodological advantages of the coupling of knowledge with a linguistic entity, such as a statement, are easy to perceive. Namely: a statement, as a linguistic entity, can always be objectified and identified with no special difficulty. With the help of a sign system, statements are observable and evident. The advantages of that can be clearly seen by thinking of a comparable situation in which someone tries to answer the question of the structures of knowledge not on the basis of statements but, on the basis of, for instance, the phenomena of consciousness. These phenomena can not be objectified and identified in the same way as statements. And they certainly can not be communicated to anyone else. So a person is never quite sure whether the other partner involved in the discussion is really oriented to the same thing as s/he herself/himself is.	The methodological advantages of the coupling of knowledge with a linguistic entity, such as a statement, are easy to perceive. Namely, a statement, as a linguistic entity, can always be objectified and identified with no special difficulty. With the help of a sign system, statements are observable and evident. The advantages of that can be clearly seen by thinking of a comparable situation in which someone tries to answer the question of the structures of knowledge not on the basis of statements but, on the basis of, for instance, the phenomena of consciousness. These phenomena can not be objectified and identified in the same way as propositional statements. As such, thus, they cannot be communicated to anyone else through propositional statements. So, a person is never quite sure whether the other partner involved in the discussion is really oriented to the same thing as s/he him/herself is.	Die methodischen Vorteile der Koppelung des Wissens an ein sprachliches Gebilde von Art der Aussage sind leicht zu sehen. Als sprachliches Gebilde läßt sich nämlich eine Aussage stets ohne besondere Mühe objektivieren und identifizieren. [...] Denn Aussagen lassen sich stets mit Hilfe eines Zeichensystems vor Augen stellen und präsent machen. Was das bedeutet, macht man sich sofort klar, wenn man sich zum Vergleich an der Situation dessen orientiert, der die Frage nach den Strukturen des Wissens nicht an Hand von Aussagen, sondern etwa an Hand von Bewußtseinsphänomenen zu beantworten unternimmt. Bewußtseinsphänomene lassen sich nämlich nicht in derselben Weise wie Aussagen objektivieren und identifizieren. Vor allem lassen sie sich nicht einem anderen mitteilen. Hier ist man daher niemals ganz sicher, ob der Diskussionspartner wirklich an derselben Sache orientiert ist wie man selbst.

epistemological underpinning" to his account of Plato's views of knowledge (Frede 1986: 464). The appropriation of passages from this chapter occurs first in N. 2002 and then again in an act of plagiarizing text recycling in N. et al. 2009.

Table 4.13 presents another selection of text appropriated from Chap. 3 of Wieland's book, and again the subject matter bears no marks of exegeting Platonic texts. The subject matter of the articles is both epistemological and linguistic, appearing sufficiently novel to persuade the editors of two specialized publications in semiotics of its value as a new contribution to that subfield of philosophy. Yet the novelty of passage is not in its content, but in its translation into English and its appearance under the name of a new author of record. Neither N. 2002 nor N. et al.

Table 4.13 Translation plagiarism and text recycling on the content and structure of sentences

N. 2002: 377	N. et al. 2009: 409	Wieland 1982: 232
One can never dissociate oneself from it in the same way one can dissociate oneself from a statement and its contents. Naturally, propositional knowledge also remains dependent on a possible conveyor, but that conveyor need not be defined individually. Therefore, one can also talk about a sentence, its contents and its structure without having to speak about the subject expressing the sentence. In a similar way, experience can certainly not be spoken of without, at the same time, taking into account an authority who is the possessor of this experience.	One can never dissociate oneself from it in the same way as one can dissociate oneself from one's own statement and its contents. To be sure, propositional knowledge also remains dependent on a possible conveyor, but such a conveyor need not be defined individually. Therefore one can talk about a sentence, its contents and its structure without having to speak about the individual expressing the sentence. By contrast, it is certainly not possible to speak about certain experience without, at the same time, taking into account the authority who is the possessor of this experience.	Man kann sich von ihr niemals so distanzieren, wie man sich von einer Behauptung und ihrem Inhalt distanzieren kann. Natürlich bleibt auch propositionales Wissen in letzter Instanz stets auf einen möglichen Träger angewiesen. Doch dieser Träger braucht nicht individuell bestimmt zu sein. Deshalb kann man von ihm auch abstrahieren. Man kann daher von einem Satz, seinem Inhalt und seiner Struktur reden, ohne zugleich von einem Subjekt reden zu müssen, das diesen Satz ausspricht oder denkt. Man kann aber nicht auf analoge Weise von Erfahrung sprechen, ohne zugleich eine Instanz in Rechnung zu stellen, die Inhaber dieser Erfahrung ist.

4.3 Case 3: The English Re-Incarnation of a German Monograph 71

Table 4.14 Translation plagiarism and duplicate publication on Socratic dialectic

N. 2002: 378–379	N. et al. 2009: 411–412	Wieland 1982: 243
The knowledge which sets Socrates apart from all his dialogue partners is not at all the type that can be represented and expressed as assertions. Rather, it is a kind of ability which proves itself able to deal with assertions, to examine them in relation to their obvious and concealed prerequisites and to refer them back to their author. This knowledge can neither be disassociated from the person nor can it be objectified. It is manifested in a capacity for theoretical and practical discernment by which it is characterized, far exceeding all its rivals.	The knowledge that sets Socrates apart from all his dialogue partners is not at all the type that can be represented and expressed as assertions. Rather, it is a kind of ability that proves itself able to deal with assertions, to examine them in relation to their obvious and concealed prerequisites and to refer them back to their author. This knowledge can neither be disassociated from the person nor can it be objectified. It is manifested in a capacity for theoretical and practical discernment by which it is characterized, far exceeding all its rivals.	Doch das Wissen, durch das sich Sokrates gegenüber allen seinen Partnern auszeichnet, ist gar nicht von der Art, daß es sich in Behauptungen darstellen und auf diese Weise mitteilbar machen ließe. Es ist von der Art einer Fähigkeit, die sich darin bewährt, mit Behauptungen umzugehen, sie auf ihre offenen und verdeckten Voraussetzungen hin zu untersuchen und sie auf ihren Urheber zurückzubeziehen. Dieses Wissen läßt sich von seiner Person nicht ablösen, und es läßt sich auch nicht objektivieren. Es zeigt sich in seiner theoretischen und in seiner praktischen Urteilskraft. Sie ist es, durch die er sich auszeichnet und alle seine Partner weit überragt.

2009 declare that they are in part translations of Wieland's monograph. Rather, they masquerade as original works of philosophy.

There are some minor differences between the two renderings of this passage from Wieland into English in N. 2002 and N. et al. 2009. The earlier of the two translates *Subjekt* with the direct cognate in English "subject" but in the later text it has been rendered as "individual." This variance, of course, does not alter the fact that both passages are linguistically respectable translations of Wieland's book.

As a last example from this instance of dispersal plagiarism, Table 4.14 presents a text from Wieland dealing with an aspect Plato's account of knowledge. The plagiarizing versions contain small additions. The word "dialogue" has been added to modify the term *Partnern* to create "dialogue partners," and the word "rather" has been introduced at the beginning of one sentence as an adverb. Again, the identity of source text and the plagiarizing versions is evident through the translation.

The plagiarizing articles by N. that appropriate text from Wieland do not appear to have impacted the downstream literature significantly. Three of the four articles have no citations according to *Google Scholar*. N. et al. 2009 has less than 20 citations.

4.4 Conclusion

This chapter has identified the phenomenon of dispersal plagiarism as a distinct form of research misconduct. This identification was supplemented by an examination of three unrelated cases from the discipline of philosophy. These cases demonstrate that dispersal plagiarism is an existing form of plagiarism that corrupts the body of published literature, and its adverse effects can persist in the downstream literature to the detriment of researchers.

Some accidental factors may have contributed to the success of these cases of dispersal plagiarism. For the first two, involving Kribbe's dissertation and Kantola's monograph, the genuine authors whose works were misappropriated did not pursue academic careers. Had they done so, perhaps they would have personally discovered the academic burgling of their work early on. Instead, the instances of plagiarism were left to be discovered after many years by third-party researchers. For the case of dispersal plagiarism involving Wieland's German monograph, the added element of having a source text in different language than the plagiarizing texts likely increased the difficulty in discovering the plagiarism.

This chapter also noted the accompaniment of duplicate publication and text recycling with acts of dispersal plagiarism. There is no necessary connection between plagiarism and inappropriate re-use of text. In the instances of dispersal plagiarism examined here, duplicate publication and text recycling greatly augmented the corruption of the body of research literature by multiplying the instances in which fraudulent texts were published.

References

[A.] 2008a. Understanding subject(s). *The Heythrop Journal* 49(3): 423–441. [Retracted in 2019 in 60(2): 339].

[A.] 2008b. A suggested basis for legal ontology. *Ratio Juris* 21 (1): 19–38. [Retracted in 2020 in 33 (1): 117].

[A.] 2008c. The illusory nature of ontological status and its implications for legal, moral and social organisation. *IJLC* 4.1: 63–77 [Retracted in 2019 in 15(1): 103].

[A.]. 2009. Transgenerational association. *IJLC* 5(2): 93–106. [Retracted in 2019 in 15 (2019): 104].

[A.] 2010. Being lucky and being deserving, and distribution. *The Heythrop Journal* 51(4): 658–669. [Retracted in 2019 in 60(2): 340].

[A.] 2011. Corporate personality. *Ratio Juris* 24(4): 471–493. [Retracted in 2020 in 33 (1): 118].

[A.] 2012. An intentional basis for corporate personality. *IJLC* 8.3: 413–430 [Retracted in 2019 in 15(1): 105].

Anonymous. 2010. *Notification*. Leuven: Leuven University Press.

Anonymous. 2017. It was brought to our attention […]. Amsterdam: John Benjamins. https://doi.org/10.1075/cvs.1.05sch.

Anonymous. 2019a. Retraction. *The Heythrop Journal* 60(2): 339.

Anonymous. 2019b. Retraction. *The Heythrop Journal* 60(2): 340.

Anonymous. 2019c. Retraction. *IJLC* 15(1): 103.

Anonymous. 2019d. Retraction. *IJLC* 15(1): 104.

Anonymous. 2019e. Retraction. *IJLC* 15(1): 105.

Anonymous. 2020a. Retraction. *Ratio Juris* 33 (1): 117.

Anonymous. 2020b. Retraction. *Ratio Juris* 33 (1): 118.

Astorri, Paolo. 2019. *Lutheran theology and contract law in early modern Germany (ca. 1520–1720)*. Paderborn: Ferdinand Schöningh.

Autor, David H. 2011. Letter. *Journal of Economic Perspectives* 25.3: 239–240. https://doi.org/10.1257/jep.25.3.239.

Barry, Bruce. 2018. Expression of concern. *Business Ethics Quarterly* 28 (2): 237–239.

Biagioli, Mario. 2012. Recycling texts or stealing time? *International Journal of Cultural Property* 19 (3): 453–476.

Dougherty, M.V., Pernille Harsting, and Russell L. Friedman. 2009. 40 cases of plagiarism. *Bulletin de Philosophie médiévale* 51: 350–391.

References

Dougherty, M.V. 2017. Correcting the scholarly record in the aftermath of plagiarism. *Metaphilosophy* 48 (3): 258–283.

Dougherty, M.V. 2018. *Correcting the scholarly record for research integrity*. Cham: Springer.

ETL. 2010. Note from the editorial board. *Ephemerides Theologicae Lovanienses* 86(1): iii. http://www.peeters-leuven.be/pdf/ETL8601EditNote.pdf.

Frede, Dorothea. 1986. Platon und die Formen des Wissens. *The Philosophical Review* 95(3): 464–467.

Halevi, Gali. 2020. Why articles in arts and humanities are being retracted? *Publishing Research Quarterly* 36 (1): 55–62.

Hallebeek, Jan. 2018. Fides integra, sana, non vaccilans. *GLOSSAE: European Journal of Legal History* 15: 223–242.

Holland, Ben. 2017. *The moral person of the state*. Cambridge: Cambridge University Press.

Hutton, Christopher. 2019. *Integrationism and the self*. New York: Routledge.

JSTOR. 2000. The origins of probabilism in late scholastic moral thought. https://www.jstor.org/stable/26170043.

Kantola, Ilkka. 1994. *Probability and moral uncertainty in late medieval and early modern times*. Helsinki: Luther-Agricola-Society.

Kribbe, Hans. 2003. *Corporate personality*. PhD. thesis, LSE. http://etheses.lse.ac.uk/2659.

Korzo, Margarita A. 2019. Probabilism and the problem of 'uncertain' conscience in the early modern times [In Russian]. *Ethical Thought* 19 (1): 63–75.

Lin, Wen-Yau Cathy. 2020. Self-plagiarism in academic journal articles. *Scientometrics* https://doi.org/10.1007/s11192-020-03373-0.

Maryks, Robert Aleksander. 2008. *Saint Cicero and the Jesuits*. Aldershot: Ashgate Publishing.

[N.]. 2002 'Know how' and 'Know that'. In *Semiotics 2001*, ed. Scott Simkins and John Deely, 371–382. New York: Legas.

[N.]. et al. 2009. Subjectivity out of irony. *Semiotica* 173: 397–416.

[N.]. 2005. Being in accordance with oneself. In *Controversies and subjectivity*, ed. Pierluigi Barrotta and Marcelo Dascal, 75–90. Amsterdam: John Benjamins.[Corrigendum issued in 2017. https://doi.org/10.1075/cvs.1.05sch].

[N.]. 2014. A controversy that never happened. In *Perspectives on theory of controversies and the ethics of communication*, ed. Dana Riesenfeld and Giovanni Scarafile, 199–208. Dordrecht: Springer.

Oransky, Ivan. 2019. Cribbing from Kribbe: UK criminology prof loses four papers for plagiarism. *Retraction Watch*. https://retractionwatch.com/2019/01/09/cribbing-from-kribbe-uk-criminology-prof-loses-four-papers-for-plagiarism.

Pine, Gregory. 2019. The incipient probabilism of Francisco de Vitoria. *Nova et Vetera, English Edition* 17 (3): 717–746.

Roig, Miguel. 2015. *Avoiding plagiarism, self-plagiarism, and other questionable writing practices*. USDHHS/ORI. https://ori.hhs.gov/sites/default/files/plagiarism.pdf.

RTPM. 2009. A note from the editorial board. *Recherches de Théologie et Philosophie médiévales* 76(2): v–vi. http://www.peeters-leuven.be/pdf/RTPM7602009.pdf.

Ryan, Eugene E. 1984. Platon und die Formen des Wissens. *Journal of the History of Philosophy* 22 (1): 115–116.

[S.] et al. 1999. Probabilism and its methods. *Ephemerides Theologicae Lovanienses* 75(4): 359–394. [Retracted in 2010 in 86 (1): iii].

[S.]. 2000. The origins of probabilism in late scholastic thought. *Recherches de Théologie et Philosophie médiévales* 67(1): 114–157. [Retracted in 2009 in 76 (2): v–vi].

[S.]. 2001. Moral psychology after 1277. In *Nach der Verurteilung von 1277*, ed. Jan A. Aertsen et al., 795–826. Berlin: Walter de Gruyter.

[S.]. 2003. Henry of Ghent on freedom and human action. In *Henry of Ghent and the transformation of scholastic thought*, ed. Guy Guldentops and Carlos Steel, 201–225. Leuven: Leuven University Press. [Expression of Concern in: Anonymous 2010].

[S.]. 2004a. Scrupulosity and conscience. In *Contexts of conscience in early modern Europe, 1500–1700*, ed. Harald E. Braun and Edward Vallance, 1–16, 182–188. New York: Palgrave Macmillan.

[S.]. 2004b. The scope and limits of moral deliberation. In *Imagination in the later Middle Ages and early modern times*, ed. Lodi Nauta and Detlev Pätzold, 35–57. Leuven: Peeters.

[S.]. 2005. Moral philosophy and the conditions of certainty. In *Metaphysics, soul, and ethics in ancient thought*, ed. Ricardo Salles, 507–550. Oxford: Clarendon Press. [Expression of Concern in: Salles 2010].

Salles, Ricardo (ed.). 2010. *Metaphysics, soul, and ethics in ancient thought*. Oxford: Clarendon Press.

Saunders, Joss. 2010. Plagiarism and the law. *Learned Publishing* 23 (4): 279–292.

Schüssler, Rudolf. 2005. On the anatomy of probabilism. In *Moral philosophy on the threshold of modernity*, ed. Jill Kraye and Risto Saarinen, 91–113. Dordrecht: Springer.

Schüßler, Rudolf. 2006. Moral self-ownership and *ius possessionis* in scholastics. In *Transformations in medieval and early-modern rights discourse*, ed. Virpi Mäkinen and Peter Korkman, 149–172. Dordrecht: Springer.

Taylor, C.C.W. 1983. Plato and the written word. *The Classical Review* 33 (1): 58–60.

Thielen, Joanna. 2018. When scholarly publishing goes awry: Educating ourselves and our patrons about retracted articles. *Portal: Libraries and the Academy* 18 (1): 183–198.

Vismara, Paola. 2018. Moral economy and the Jesuits. *Journal of Jesuit Studies* 5: 610–630.

Wieland, Wolfgang. 1982. *Platon und die Formem des Wissens*. Göttingen: Vandenhoeck and Ruprecht.

Wigley, Simon. 1998. *The role of desert in distributive justice*. PhD. thesis, LSE. http://etheses.lse.ac.uk/1504.

Chapter 5
Magisterial Plagiarism

Abstract Could plagiarism cause more harm when perpetrated in some academic disciplines than in others? This chapter considers an affirmative answer, examining the effects of a particular form of disguised plagiarism committed in the specific disciplinary context of Catholic theology. Members of the Catholic hierarchy regularly produce texts considered to be authoritative for the faithful. These magisterial texts, authored by popes, cardinals, and bishops, shape the contours of Catholic theology in its various subfields, including work done in systematic, biblical, and historical theology. Plagiarism corrupts magisterial texts in two ways. First, when ghostwriters for members of the Catholic hierarchy secretly turn in plagiarizing work to their unsuspecting employers, the subsequent promulgation of these defective texts by members of the Catholic hierarchy harms the quality of magisterial teaching that informs the practice of theology. Second, when theologians themselves plagiarize magisterial texts in their own theological writings, falsely presenting the words of popes, cardinals, and bishops as their own, the prior magisterial endorsement of the source texts is concealed to readers.

Keywords Ghostwriters · Plagiarism · Theology · Authority · Tradition · Magisterium

The practice of Catholic theology is shaped to no small degree by the reception of magisterial texts issued by popes, cardinals, and bishops. On a traditional view, the term *magisterium* refers to the teaching authority that members of the Catholic hierarchy possess in order to defend and exposit the truths of faith. Popes, cardinals, and bishops typically exercise their magisterial authority by issuing a variety of texts. Papal encyclicals, apostolic exhortations, homilies, and various formal addresses, as well as pastoral letters by cardinals or bishops, are generally recognized to be among magisterial texts possessing varying degrees of weight (see Dulles 2007; Sullivan 1983). The Catechism and the Second Vatican Council have each emphasized the special magisterial authority possessed by the pope and by bishops in hierarchical communion with the pope (*Catechism of the Catholic Church* 1995: §§85–100; §§890–892; §§2033–2034). The status of magisterial teaching is expressed in detail in the Dogmatic Constitution of the Church promulgated by Pope Paul VI at the Second Vatican Council. The Council stated:

For bishops are [...] teachers endowed with the authority of Christ, who preach to the people committed to them the faith they must believe and put into practice. [...] In matters of faith and morals, the bishops speak in the name of Christ and the faithful are to accept their teaching and adhere to it with a religious assent. (*Lumen Gentium* 1964: §25).

The various allocutions by prelates in their capacity of teachers of the faith therefore possess an unrivaled authority for those in the Catholic tradition. Interpreting them is a large part of Catholic theology (Sullivan 1996).

The distinctive role of magisterial texts in the practice of Catholic theology provides an illuminating way of approaching the pernicious effects of plagiarism. Some members of the church hierarchy employ ghostwriters to assist in the writing of magisterial texts. The practice is not in itself controversial. When those ecclesiastical ghostwriters plagiarize, however, the magisterial texts that are subsequently authorized and promulgated by their employers are defective. These plagiarizing magisterial texts negatively influence the practice of Catholic theology. This chapter examines specific instances where plagiarism has corrupted the production of magisterial texts and their reception. Special consideration is given to the work of plagiarizing ghostwriters who have assisted two cardinals. Furthermore, when theologians, in their capacity as theologians, themselves plagiarize by presenting previously promulgated magisterial texts in the guise of their own new theological writing, a different kind of corruption of the discipline occurs. Readers of those plagiarizing theological texts encounter magisterial documents stripped of their magisterial endorsement, and this privation impedes a proper assessment of their quality.

5.1 Cardinal William Levada's Ghostwriter

In 2007, The American College in Louvain, Belgium, a now-closed American seminary founded in 1857, celebrated the sesquicentennial of its founding. The year-long schedule of festivities included a mass on March 17th and the homilist was American Cardinal William Levada. Formerly the Archbishop of San Francisco, Levada at the time was holding the office of Prefect of the Congregation for the Doctrine of the Faith (CDF), the important curial congregation responsible for safeguarding Catholic teaching.

A few days after the mass, the text of Levada's homily appeared on the website of The American College, and it received a positive reception in the Catholic press (The American College 2007; Palmo 2007). The text was then published as an article in *Origins* early the next month, appearing in the April 5 edition, where it was titled "The Liturgical Foundations of Theology" (Levada 2007). In its published form as an article in *Origins*, Levada's text gives every indication of employing proper attribution to sources, since it contains quotations and in-text citations to various works throughout. Despite these attributions, however, much of the article is not original.

Levada's unidentified ghostwriter produced an apparently plagiarizing text for the cardinal that incorporates lengthy sections of an unreferenced scholarly theological

5.1 Cardinal William Levada's Ghostwriter

article by Benedictine theologian and priest (now abbot) Jeremy Driscoll. Much of Levada's article is a re-appropriation—without attribution of any kind—of the previously published article by Driscoll. The article by Driscoll first appeared in the liturgical journal *Antiphon* in 2000 with the title "The Fathers and Eucharistic Preaching" (Driscoll 2000) and then again in 2002 in the ecumenical theology journal *Pro Ecclesia* with the lengthier title "Preaching in the Context of the Eucharist: A Patristic Perspective" (Driscoll 2002). The 2002 version was then published the next year as Chapter 10 of the monograph *Theology at the Eucharistic Table* (Driscoll 2003: 215–236). No reader of Levada's article, however, can judge from the text itself that a third of it derives from Driscoll's often-published work. The unidentified ghostwriter has employed some subtle manipulations to disguise the plagiarism that corrupts Cardinal Levada's magisterial text.

5.1.1 A Subtle Change in Subject Matter

As its original titles suggest, the article by Driscoll concerns preaching, and it considers Patristic approaches to the act of preaching in the context of the celebration of the Eucharist. Table 5.1 presents a selection from Driscoll's text, with the corresponding unreferenced version of the text as it appears in Levada's plagiarizing article. The overlap between the two is highlighted.

The major change in the plagiarizing version produced by Levada's ghostwriter occurs at the beginning of the passage, with the substitution of the expression "sound theology" for the original expression "good preaching." This modification is remarkable, since the two expressions are not synonymous in the discipline of theology; Driscoll's text has been bent by the plagiarizing ghostwriter to a different purpose. The other modifications made to the passage are comparatively minor.

Due to the copying by Levada's plagiarizing ghostwriter, Driscoll's words have been incorporated into a text promulgated by a cardinal, none other than the Prefect of the CDF. The cardinal's endorsement of sentences and paragraphs originally found in

Table 5.1 A change in subject matter from Driscoll 2000/2002/2003 to Levada 2007

[Cardinal Levada's Ghostwriter] Levada 2007: 688	Driscoll 2000: 30 / 2002: 27 / 2003: 219
Sound theology must derive from an awareness of what wonderful mysteries are taking place during the celebration of the liturgy. These wonders begin with the proclamation of the word. The word of God recalls the wonderful deeds of God in the history of salvation. But this is not a question of mere memory. Whenever it is proclaimed, the word of God becomes a new communication of salvation for those who hear it. The event from the past that is proclaimed becomes "event" for the listening assembly. And ultimately all the events of Scripture merge into the one event that encompasses them all — namely, Christ in the hour of his paschal mystery. The moment of listening to the word in the liturgy — the word which proclaims ultimately the Lord's death and resurrection — becomes in the very hearing an event of salvation for those who listen, nothing less than the same event which the words proclaim.	Good preaching must derive from an awareness of what wonderful mysteries are taking place in the celebration of the liturgy. These wonders begin with the proclamation of the Word. The Word of God recalls the wonderful deeds of God in the history of salvation. But this is not a question of mere memory. Whenever it is proclaimed, the Word of God becomes a new communication of salvation for those who hear it. The event proclaimed becomes the event of the listening assembly, and ultimately all the events of Scripture are reducible to one event which encompasses them all; namely, Christ in the hour of his Paschal Mystery. The moment of listening to the Word — the Word which proclaims ultimately the Lord's death and resurrection — becomes in the very hearing an event of salvation, nothing less than the same event which the words proclaim.

Table 5.2 From preaching to theology

[Cardinal Levada's Ghostwriter] Levada 2007: 688	Driscoll 2000: 31 / 2002: 28 / 2003: 220–221
Theology must speak of these things. Theology exists because of these things. The theologian must be capable of understanding them more deeply, explaining them, defending them against error, indicating ways to proclaim them, lifting the community's minds and hearts up toward them. The Scriptures must be expounded in this way and not left at the level of an exegetical exercise which explains the text only in its original historical context. All the texts must be brought to the event that encompasses them: the Lord's death and resurrection. That through the Eucharist about to be celebrated we have communion in the very same death and resurrection — this too must be proclaimed and explained.	Preaching during the Eucharist must speak of these things. The preacher must be capable of explaining them, proclaiming them, lifting the community's minds and hearts up toward them. The Scriptures must be expounded in this way, and not left at the level of an exegetical exercise which explains the text only in its original historical context. All the texts must be brought to the event that encompasses them: the Lord's death and resurrection. That through the Eucharist about to be celebrated we have communion in the very same death and resurrection— this too must be proclaimed and explained.

a theology journal article is not negligible. On a traditional Catholic view, Driscoll's theological writing is receiving a major "upgrade" by being transformed from merely a theological work to now a magisterial one. That is, Driscoll's words are appearing anew in a text in which Levada is exercising his teaching function in the church. What was merely academic or scholarly has now also become magisterial with this new attestation by the cardinal.

The ghostwriter's substitution of the topic of preaching for the topic of theology continues throughout Levada's article. Table 5.2 shows how the expressions "preaching" and "the preacher" have been replaced by "theology" and "the theologian" in the plagiarizing version.

With this substitution, Driscoll's description of the role of a preacher who exposits the scriptures in the context of a Catholic liturgy has been modified by Levada's ghostwriter to form a description of the role of a theologian who exposits the scriptures in the practice of academic theology. This alteration changes the overarching emphasis of the text, but the simple switch does not result in the production of a genuine, original, authentic work. Since the key terms *preaching* and *theology* do not have coextensive meanings in the discipline of theology, one must ask whether the ghostwriter's plagiarizing construction is still intelligible. If one answers in the affirmative, one renders oneself vulnerable to the *ex hypothesi* objection that theology must lack the rigor found in other disciplines if a coherent contribution can be made by taking an article and simply substituting one discrete technical term from the discipline for another. Both the source article and the plagiarizing article contain systematic, biblical, and historical claims, and it is difficult to conceive how these claims could remain reliable when migrating to different contexts.[1]

Table 5.3 exhibits how a particular historical claim made by Driscoll re-appears in new guise in the plagiarizing version. The passage in its original version concerns the view of the purpose of preaching during the Patristic period of early Christianity.

---

[1] The substitution of key terms exhibited in Tables 5.1–5.2 are instances of the phenomenon of *template plagiarism* (see Chap. 7). Driscoll's text is being used a template, and key terms are being replaced with new ones to give the illusion of new scholarship.

5.1 Cardinal William Levada's Ghostwriter

Table 5.3 Driscoll on the function of Patristic preaching

[Cardinal Levada's Ghostwriter] Levada 2007: 689	Driscoll 2000: 32 / 2002: 31 / 2003: 224
The golden age of patristic preaching was between Nicea and Chalcedon precisely because doctrine, metaphysically grounded, was known to be of vital importance for the salvation of the Christian people. Again, this was not mere academic talk. The Fathers used the Scriptures and the celebration of the sacraments as an occasion to develop this or that necessary doctrinal emphasis to clearly oppose some false idea, going to the roots of the problem through their refined metaphysic. The most important doctrines remain the same through the ages and need to be approached again and again by theology and in our preaching — namely, the divine and human natures of Christ; their union in the divine person of the Son; and the mystery of the Holy Trinity which Christ reveals in his paschal mystery.	The golden age of patristic preaching was between Nicea and Chalcedon precisely because doctrine was known to be important and comes into most of the sermons. Again, this was not academic talk. The Fathers used the Scriptures and the celebration of the sacraments as an occasion to develop this or that necessary emphasis to oppose clearly some false idea. [...] The most important doctrines remain the same through the ages and need to be approached again and again in preaching; namely, the divine and human natures of Christ and the mystery of the Trinity which Christ reveals in his Paschal Mystery.

Toward the end of the passage, the expression "by theology" has been added in the plagiarizing version in the service of bending Driscoll's text to its new ostensible subject matter; the passage in its revised version now discusses preaching *and* theology.

Deserving of mention is the fact that the second half of the passage from Driscoll presented in Table 5.3 is excerpted—this time with quotation marks and a full accurate citation—in a different talk that Levada presented a year later at a theology conference at the University of Notre Dame. In his address there, later published both in *Origins* and on the Vatican website, Levada explicitly praised Driscoll for "important insight in commenting on the model provided by the Fathers in the patristic age" (Levada 2008a: 606, 2008b). Levada also discusses—with quotations—two other works by Driscoll in that address. One must conclude that Levada's ghostwriter is conversant with Driscoll's extensive body of theological publications. In the specific instance of the 2007 plagiarizing article, however, Levada's ghostwriter has concealed the substantial dependency of the article on Driscoll's work.

5.1.2 First-Person Plagiarism

The plagiarism in Levada 2007 extends to the appropriation of claims that Driscoll originally expressed in the first person. Such repetition is remarkable, not simply because Driscoll and Levada are different persons, but because the endorsement of a position by a theologian engaged in the act of theological analysis is fundamentally different in kind from the endorsement of a position by a member of the church hierarchy who is engaged in the exercise of his teaching authority. Only the latter act is considered in the Catholic tradition to have a distinctive magisterial quality. One theologian has expressed the difference succinctly, stating, "theologians do not, as such, have a magisterial status" (Dulles 2007: 39). Making the same point about magisterial teaching, another theologian explains that "there is a genuine kind of teaching in the church which is distinct from and different from the work of the university professor of theology" (Sullivan 1983: 45). Table 5.4 exhibits a passage where Levada's ghostwriter plagiarizes a passage from Driscoll written in the first person.

Table 5.4 Plagiarism and the use of the first person

[Cardinal Levada's Ghostwriter] Levada 2007: 689	Driscoll 2000: 32 / 2002: 32 / 2003: 224–225
I am not suggesting that theologians and preachers ought simply to stand up and talk more about these things. Rather, I am drawing our attention once again to the fact that these doctrines are the deepest sense of what the Scriptures proclaim and that this deepest sense was discovered precisely when the Scriptures were proclaimed in the liturgical assembly and when the Scriptures became sacrament in the eucharistic rite.	I am not suggesting that preachers ought simply to stand up and talk more about these things. Rather, I am claiming that these doctrines are the deepest sense of the Scriptures and that this deepest sense was discovered precisely when the Scriptures were proclaimed in the liturgical assembly and when the Scriptures became sacrament in the eucharistic rite.

Again, the text has been modified so that "preachers" is expanded to "theologians and preachers."

The personal pronoun "I" as it appears in Driscoll's text is the "I" of a particular theologian in the exercise of his capacity as theologian. As it appears in the plagiarizing 2007 article, however, the "I" is the "I" of the Prefect of the CDF in the exercise of his teaching authority. The magisterial quality possessed by the later formulation of the text is absent from the earlier formulation, even though the words being endorsed in the original text and in the plagiarizing version are verbatim and near-verbatim.

It would be incorrect to assume, of course, that all pronouncements made by members of the Roman Catholic hierarchy on theological matters are intrinsically magisterial. Popes, cardinals, and bishops can set aside their teaching authority and produce non-magisterial works that are simply theological. In 2007, for example, Pope Benedict XVI did so when he published the first volume of his three-part study *Jesus of Nazareth*. In the foreward that begins the book, the pope specified clearly that, "this book is in no way an exercise of the magisterium, but is solely an expression of my personal search for 'the face of the Lord'." He then added, "Everyone is free, then, to contradict me" (Ratzinger 2007: xxiii–xxiv). There is no evidence that Levada had set aside the special teaching authority he possesses in virtue of his episcopal consecration when he presented "The Liturgical Foundations of Theology." To the contrary, its initial promulgation in the context of a mass, prior to its publication in *Origins*, suggests that the document is an exercise of his teaching authority.

Tables 5.1–5.4 have presented select passages from Levada's article that derive from one uncited source, Driscoll 2000, 2002, 2003. The passages offered here are representative, rather than exhaustive. As noted above, no less than one third of the content of Levada's nearly 2500-word article is taken from this one unacknowledged source text. Published acts of plagiarism in theology corrupt the quality of later theological work by others, and this case is no exception. It should not be surprising that a published article ostensibly authored by the Prefect of the CDF (and certainly authorized by him) that concerns the nature of theology would be of interest to Catholic theologians. The detrimental effect of the plagiarism by Levada's ghostwriter is manifested, for example, in a theology dissertation completed in 2008 at an American Catholic university (Mele 2008: 66). The dissertation engages Levada's article, citing two passages from it. Both passages really derive from Driscoll's text, however. The dissertation writer undoubtedly believes he is engaging with the thought of the Prefect of the CDF on the "liturgical foundations of theology," but in fact he is unwittingly engaging with the thought of the theologian Driscoll on "preaching in the context of the Eucharist."

5.2 Cardinal Marc Ouellet's Ghostwriter

Just over 5 weeks after Cardinal Levada spoke at The American College, and 2 weeks after the text was published in *Origins*, Cardinal Marc Ouellet was on the other side of the globe giving a lecture at Assumption University in Windsor, Ontario. At the time, the university also happened to be celebrating its sesquicentennial anniversary. The occasion of the lecture was the bestowal on Ouellet of the university's Christian Culture Gold Medal Award. Ouellet's talk, delivered on April 22, 2007, was titled "A Culture of the Eucharist for a Civilization of Love." The then-Archbishop of Quebec and Primate of Canada spent much of his address discussing the Eucharist and describing an upcoming Eucharistic Congress. A month and a half later, the text of Ouellet's address was published in *Origins* (Ouellet 2007).

At least of the third of the talk is unoriginal, however. Cardinal Ouellet—like Cardinal Levada—apparently employs a plagiarizing ghostwriter. In this case, Ouellet's plagiarizing ghostwriter misappropriated passages without attribution from a wide range of works to produce a fraudulent amalgam for the cardinal's address.

Chief among the works appropriated by the plagiarizing ghostwriter for Ouellet was none other than the text of Cardinal Levada's earlier talk presented at The American College the previous month. Given the substantial dependency of the text of Ouellet 2007 upon Levada 2007, the *terminus post quem* for the work of Ouellet's ghostwriter is the date of Levada's address at The American College (March 17, 2007), and the *terminus ante quem* is the date of Ouellet's address at Assumption University (April 22, 2007). It appears that Ouellet's ghostwriter worked within a very narrow timeframe.

5.2.1 Double Plagiarism

Since, as noted above, Levada himself had also used a plagiarizing ghostwriter, one finds herein the remarkably complex phenomenon of a plagiarist plagiarizing a plagiarizing text produced by a different plagiarist. The case of double plagiarism is also remarkable insofar as one magisterial text endorsed by one cardinal is being plagiarized by another magisterial text endorsed by another cardinal. As shall be seen, however, these facts are not the strangest elements of this situation.

Table 5.5 presents a passage from Ouellet's article alongside Levada's article with the overlap highlighted. The ellipses in the right column indicate that Ouellet's ghostwriter has condensed a lengthier section of Levada 2007 in producing this portion of Ouellet 2007. (See Chap. 3 for a discussion of the phenomenon of compression

Table 5.5 The presence of Driscoll's text in Levada 2007 and Ouellet 2007

[Cardinal Ouellet's Ghostwriter] Ouellet 2007: 62	[Cardinal Levada's Ghostwriter] Levada 2007: 688–689
Authentic lay formation and sound theology <u>must derive from an awareness of what wonderful mysteries are taking place</u> during <u>the celebration of the liturgy. These wonders begin with the proclamation of the word</u> <u>of God</u> that <u>recalls the wonderful deeds of God in the history of salvation. But this is not a question of mere memory. Whenever it is proclaimed, the word of God becomes a new communication of salvation for those who hear it.</u> All the words of the law and the prophets become concrete reality in the Word made flesh. <u>Echoing</u> this <u>pattern, the scriptural words proclaimed in the liturgy become sacrament; that is, the ritual actions and words performed around the community's gifts of bread and wine proclaim in their own way and at an even deeper level</u> — at a more concrete level — <u>the one and only event of salvation: the Lord's death and resurrection.</u> Theology and pastoral formation <u>must speak of these things.</u> Theology exists because of these things. <u>All the texts must be brought to the event that encompasses them: the Lord's death and resurrection. The most important doctrines remain the same through the ages and need to be approached again and again</u> by theology and <u>in</u> our <u>preaching; namely, the divine and human natures of Christ;</u> their union in the divine person of the Son; <u>and the mystery of the</u> Holy <u>Trinity</u> that <u>Christ reveals in his paschal mystery. These doctrines are the deepest sense of</u> what <u>the Scriptures</u> <u>proclaim</u> and <u>that this deepest sense was discovered precisely when the Scriptures were proclaimed in the liturgical assembly and when the Scriptures became sacrament in the eucharistic rite.</u>	Sound theology <u>must derive from an awareness of what wonderful mysteries are taking place</u> during <u>the celebration of the liturgy. These wonders begin with the proclamation of the word.</u> The word of God recalls the wonderful deeds of God in the history of salvation. But this is not a question of mere memory. Whenever it is proclaimed, the word of God becomes a new communication of salvation for those who hear it. [...] All the words of the law and the prophets become concrete reality in the Word made flesh. <u>Echoing</u> this <u>pattern, the scriptural words proclaimed in the liturgy become sacrament; that is, the ritual actions and words performed around the community's gifts of bread and wine proclaim in their own way and at an even deeper level</u> — at a more concrete level — <u>the one and only event of salvation: the Lord's death and resurrection.</u> [...] Theology <u>must speak of these things.</u> Theology exists because of these things. [...] <u>All the texts must be brought to the event that encompasses them: the Lord's death and resurrection.</u> [...] <u>The most important doctrines remain the same through the ages and need to be approached again and again</u> by theology and <u>in our preaching — namely, the divine and human natures of Christ; their union in the divine person of the Son; and the mystery of the Holy Trinity which Christ reveals in his paschal mystery.</u> [...] <u>these doctrines are the deepest sense of</u> what <u>the Scriptures</u> proclaim <u>and that this deepest sense was discovered precisely when the Scriptures were proclaimed in the liturgical assembly and when the Scriptures became sacrament in the eucharistic rite.</u>

plagiarism.) There is little in the passage from Ouellet's talk that is not found in Levada's talk. The only significant addition is that where Levada's version has the expressions "sound theology" and "theology," Ouellet's version has expanded these expressions to "authentic lay formation and sound theology" and "theology and pastoral formation." One will recall, however, that a third of Levada 2007 is a plagiarism of Jeremy Driscoll's theology article. Since Ouellet 2007 is heavily plagiarizing Levada 2007, Driscoll's words not only reappear in Levada's talk but also re-reappear in Ouellet's talk. The concealed text of Driscoll in both plagiarizing works is indicated with underlining.

Driscoll's words, which were first conscripted into plagiarizing service by Levada's ghostwriter in Levada 2007, have been misappropriated again by Ouellet's ghostwriter in Ouellet 2007. The words are misappropriated not directly, but through the medium of Levada's talk, which Ouellet's plagiarizing ghostwriter has mined as a source text. The highlighted but not underlined portions in common to both columns in Table 5.5 are portions of text common to Levada 2007 and Ouellet 2007 that do not originate in Driscoll; the presence of the highlighted but not underlined text shows that Ouellet's ghostwriter is working from Levada's text rather than Driscoll directly. The extensive underlining in Table 5.5 shows clearly that Driscoll's text is nevertheless substantively persisting through two separate rounds of plagiarism performed by different ghostwriters for different cardinals. Ouellet's plagiarizing ghostwriter may not have been aware that the main source text of Levada 2007 is itself a plagiarism of Driscoll's theological work.

5.2 Cardinal Marc Ouellet's Ghostwriter

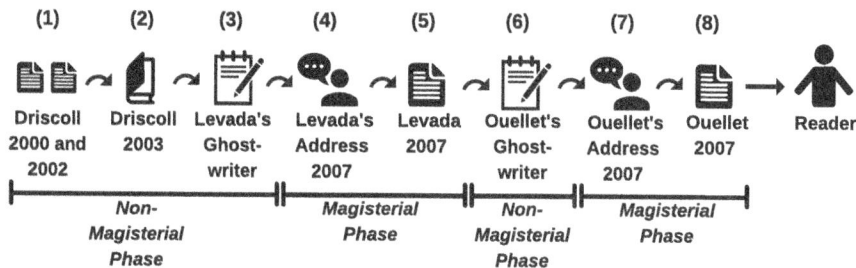

Fig. 5.1 The textual transmission of Driscoll's text

The curious result of this double plagiarism is that text from Driscoll's theological article is now part of the published output of two Roman Catholic Cardinals. To put it another way: Driscoll's words have acquired a magisterial endorsement they previously did not enjoy, and this endorsement has occurred not just once but twice. The journey of Driscoll's words from a theological journal article originally published in 2000 to two magisterial iterations in early 2007 is a complex one, involving two plagiarizing ghostwriters, two high-ranking members of the Roman Catholic hierarchy, and plagiarizing promulgations in the same publication—*Origins*—only weeks apart. The very short timeframe indicates that Ouellet's ghostwriter must have constructed the plagiarizing text from Levada 2007 within the early weeks of Levada's homily and its publication in *Origins*.

A reader who encounters the detailed discussion of the Eucharist in Ouellet 2007 likely believes she or he is reading an original document authored (or at least authorized) by Ouellet in his capacity as an authoritative teacher of the Catholic faith. But, as demonstrated above, it is not so. Figure 5.1 summarizes the complex hidden textual transmission of Driscoll's words.

Driscoll's words first appeared in print in the 2000 and 2002 as journal articles in *Antiphon* and *Pro Ecclesia* (1), and then again as a book chapter in 2003 (2). Levada's plagiarizing ghostwriter takes Driscoll's words without attribution in producing the text of Levada's address (3), which is then presented by Levada (4) to the community at The American College on March 17, 2007. The text is then published as an article in *Origins* (5). This article in *Origins* is then taken by Ouellet's ghostwriter to produce the text for Ouellet (6), and then Ouellet presents it at Assumption University on April 22, 2007 (7). Ouellet's talk is published in *Origins* as Ouellet 2007 (8), where it is encountered by readers either in print form or its online electronic format. A reader of Ouellet's 2007 article is unlikely to be aware of this exotic textual transmission and is also unlikely to be aware that many of its words originate in a theology article written by theologian Driscoll more than a half decade earlier. Further supporting the illusion of textual authenticity for the reader is the fact that Ouellet 2007 gives every appearance of following the customary forms of attribution found in academic writing. There are in-text citations and quotation marks for a variety of other texts.

Figure 5.1 also illustrates that not all iterations of Driscoll's words in the various stages of the textual transmission enjoy a magisterial endorsement; that endorsement is present only when the text is promulgated by Cardinal Levada in his talk (4) and its subsequent publication (5), and promulgated by Cardinal Ouellet in his talk (7) and its subsequent publication (8). Lacking episcopal consecration, Driscoll himself does not possess magisterial authority, and presumably the unnamed plagiarizing ghostwriters for the two cardinals do not possess it either.

One detail of this textual transmission that remains unknown is which of three versions of Driscoll's work was used by Levada's ghostwriter in compiling Levada's address. As noted above, Driscoll published the text as an article twice (Driscoll 2000, 2002), and then published it once more as a book chapter (Driscoll 2003). In another oddity, the publication of Driscoll (2002) in the journal *Pro Ecclesia* was an act of undisclosed duplicate publication; no mention of the prior publication of the text in *Antiphon* was given the later article version. The book chapter, however, references both earlier versions in an appendix titled "Reference to Original Publications (2003: 245–246). In August 2019, I wrote to the editor of *Pro Ecclesia* to inquire whether the duplicate publication was consistent with the journal's stated policy on original content. In response, the journal published a brief corrigendum the following month that references the earlier publication in *Antiphon* and declares, "A reference to a paper previously published by the same author should have been included" (Driscoll 2019: 1).

The appearance of both plagiarizing magisterial works—Levada 2007 and Ouellet 2007—as defective articles in the same publication might lead one to wonder whether the editors of *Origins* might bear any responsibility in the proliferation of defective plagiarizing articles. Even if one considers that the editors are not culpable for having published the earlier plagiarizing article by Levada, one may wonder why they failed to recognize—only weeks later—that Ouellet's article overlapped with it to a significant degree. One mitigating factor, as chronicled above, may be that Driscoll's words were modified as they travelled through the two iterations. A key transformation of Driscoll's text was the substitution by Levada's ghostwriter of the expression "sound theology" for "preaching." Next, Ouellet's ghostwriter expanded Levada's ghostwriter's modified expression "sound theology" to "authentic lay formation and sound theology." Figure 5.2 illustrates these manipulations to Driscoll's words by the two plagiarizing ghostwriters.

As the underlined terms in Fig. 5.2 are not synonymous, one must wonder about the quality of the content of the two later articles in *Origins*. The terms *preaching*, *theology*, and *lay formation* are indisputably important ones in theology, each with a long history of careful reflection by theologians, and they each are presumably understood in precise ways in theological contexts. The substitutions found in Levada 2007 and Ouellet 2007 call into question the intelligibility of the texts manufactured by the two plagiarizing ghostwriters. Have they each produced coherent works of Catholic teaching, or are the plagiarizing documents simply theological word-salads?

The ghostwriter for Ouellet did plagiarize a portion of Levada 2007 that Levada's ghostwriter did not take from Driscoll. Levada's text contains some ostensibly autobiographical reflections that are expressed in the first person. These apparently autobiographical reflections have been slightly adapted by Ouellet's ghostwriter, no doubt partly for the reason of making Levada's spoken words fit the new occasion in

5.2 Cardinal Marc Ouellet's Ghostwriter

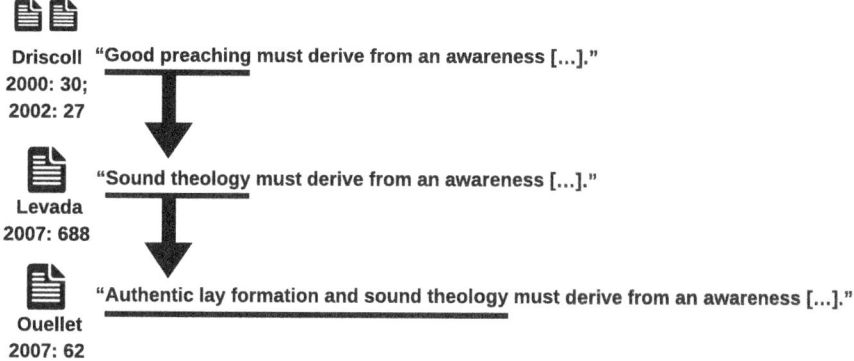

Fig. 5.2 The substitution of key terms by two plagiarizing ghostwriters

Table 5.6 The institutional hopes of two cardinals

[Cardinal Ouellet's Ghostwriter] Ouellet 2007: 62	[Cardinal Levada's Ghostwriter] Levada 2007: 688
It is my fervent hope, and the hope of the Holy Father and the bishops of this country — for this Catholic university and indeed for the whole church — that doctrine might become a more vital and active dimension of your educational ministry — not doctrine merely as this is usefully discussed in speculative schools and learned writing, but rather this doctrine brought to life by continual nourishment from its eucharistic source, this doctrine as the precise and beautiful formulation of the deepest sense of what is happening in the liturgical assembly when the word of God is proclaimed and the eucharistic rites are celebrated.	It would be my hope for this seminary and indeed for the whole church that doctrine might become a more vital and active dimension of a priest's ministry and preaching — not doctrine merely as this is usefully discussed in speculative schools and learned writing, but rather this doctrine brought to life by continual nourishment from its eucharistic source, this doctrine as the precise and beautiful formulation of the deepest sense of what is happening in the liturgical assembly when the word of God is proclaimed and the eucharistic rites are celebrated.

which they were to be presented by Ouellet. Table 5.6 presents a passage expressed in the first person that is common to both works but not derived Driscoll's work.

Levada's autobiographical expression of hope has been amplified by Ouellet's plagiarizing ghostwriter in four significant ways to form Ouellet's autobiographical expression of hope. First, the expression "Catholic university" has been substituted for "seminary" to reflect the move from The American College seminary in Belgium to Assumption University in Canada. Second, Levada's words about "a priest's ministry and preaching" have been changed to "your education ministry," since Ouellet's audience is primarily university students and lay professors rather than seminarians and priests. Third, Ouellet's text adds "the Holy Father and the bishops of this country" as a select group also expressing hope with Ouellet. And lastly, Ouellet's hope has become "fervent."

5.2.2 Other Sources Appropriated by Ouellet's Ghostwriter

Levada's ghostwriter and Ouellet's ghostwriter are likely non-identical. The evidence includes the differing plagiarism methodologies they employ, the different

Table 5.7 Ouellet's ghostwriter's use of Schillebeeckx

[Cardinal Ouellet's Ghostwriter] Ouellet 2007: 62	Edward Schillebeeckx 1990: 78
Through the Eucharist we are challenged at the level of our history to realize as much as possible what we celebrate sacramentally: bread for all, salvation and liberation for all, including greater respect for creation and promotion of a safer environment. The eucharistic Christ is really present as bread for the poor. The Christ who is sacramentally present is a constant reminder and pledge that our sharing of the bread is not in vain. In order to keep the eucharistic reality credible, we have to devote ourselves to a better, more just and habitable world.	Precisely through the eucharist we are challenged at the level of our history to realize as much as possible of what we celebrate sacramentally: bread for all, salvation and liberation for all. The eucharistic Christ is really present as bread for the poor. The Christ who is sacramentally present stands surety that our sharing of the bread, too, is not in vain. [...] And in order to keep this eucharistic reality credible, we have to devote ourselves to a better, juster world.

nationalities of the two cardinals, and, presumably, a difference in their workplace countries (Vatican City; Canada). Levada's ghostwriter appears to have used only one source—Driscoll 2000/2002/2003—and drew heavily from it in manufacturing the text of Levada 2007. In contrast, Ouellet's ghostwriter employs a patchwork plagiarism methodology, drawing passages from several sources, only one of which is Levada 2007.

Table 5.7 displays a short portion of text from Ouellet 2007 that plagiarizes an English translation of Edward Schillebeeckx's *For the Sake of the Gospel*, a book that is not referenced in Ouellet 2007 (Schillebeeckx 1990).

An influential Dominican theologian who assisted attending Dutch bishops at the Second Vatican Council, Schillebeeckx himself was rumored to be a ghostwriter—but not a plagiarizing one—for a pastoral issued by Dutch bishops in the advent of the council (Schoof 2014: xxi). Schillebeeckx periodically had some of his theological writings investigated for unorthodoxy by the CDF, and it is notable that some of his words here are acquiring a magisterial quality via their promulgation by Cardinal Ouellet. Readers of Ouellet 2007 are certainly unlikely to suspect that they are encountering the work of Schillebeeckx.

Among the various works misappropriated by Ouellet's ghostwriter, Levada 2007 is a major undisclosed source text, but it is not the one most extensively used in Ouellet 2007. The work apparently most favored there by Levada's plagiarizing ghostwriter is an early 2005 conference paper by theologian Igor Kowalewski. Published online on the website of the Vatican Congregation for Clergy with the title "The Spiritual Development of Christians and of the Church in Light of Pope John Paul II's Apostolic Letter *Mane nobiscum domine*," the occasion for the paper was a conference organized by the Vatican congregation. Passages taken from Kowalewski 2005 account for more than 15% of the total text of Ouellet 2007.

5.3 A Second Production by Ouellet's Plagiarizing Ghostwriter

Ouellet 2007 is not the only plagiarizing compilation produced by the ghostwriter for the cardinal. On May 30, 2008, Cardinal Ouellet presented the keynote address at the International Catholic Media Convention held in Toronto, and the paper was

5.3 A Second Production by Ouellet's Plagiarizing Ghostwriter

titled "The New Evangelization and the Mass Media." The text of the address was quickly published online and then the following month appeared as an article in *Origins* (Harrison 2008; Ouellet 2008a). It is also a plagiarizing compilation that exhibits the same patchwork-style plagiarism whereby passages from the writings of others are fashioned to produce the illusion of a unified text of new content. Again, in some places Ouellet's plagiarizing ghostwriter includes quotation marks and in-text citations for select works, so the use of these conventions generates the appearance that sources are being credited in the customary way of academic writing. The use of these conventions also supports the illusion that texts that are not designated as originating elsewhere are original to Ouellet.

Previously promulgated magisterial texts written by others are the principal source texts that constitute Ouellet 2008a. One might be tempted to judge that the ghostwriter has been bold in selecting passages from various magisterial sources. One major source is a 2000 address by Cardinal Joseph Ratzinger that also has a somewhat similar-sounding title to the plagiarizing article, as it is titled "The New Evangelization: Address to Catechists and Religion Teachers." At the time Ouellet's ghostwriter was cobbling together 2008a, Ratzinger was exercising the office of the papacy as Pope Benedict XVI. Table 5.8 exhibits the overlap between a portion of Ouellet 2008a and the 2000 text by the man who would become pope.

Ouellet 2008a is a compilation that largely appropriates without credit previously promulgated and published magisterial texts. The use of a work like Ratzinger 2000 is not an outlier; Ouellet's ghostwriter also incorporates, without citation, a portion of *Gaudium et spes*, the Pastoral Constitution of the Church from the Second Vatican Council (Ouellet 2008a: 94, column b, lines 4–12 = *Gaudium et spes* §92).

The manipulation of magisterial sources by Ouellet's plagiarizing ghostwriter is at times subtle. Table 5.9 discloses that a short passage of Ouellet 2008a derives from an uncredited brief issued by the Vatican Information Service (VIS) of the Holy See Press Office. In its original formulation in VIS 2008, the passage contains (1) a quotation by Archbishop Claudio Maria Celli, the president of the Pontifical Council for Social Communications; (2) a quotation within Celli's quotation from Benedict XVI's statement for World Communications Day; and (3) some commentary by the unnamed writer of the news brief. In manipulating the VIS brief to construct Ouellet's address, the plagiarizing ghostwriter has removed any references to both the archbishop and the pope, taking their words verbatim and near verbatim, and also taking verbatim text from the VIS brief itself. Audience members of Ouellet's address, as well as readers of the published version in Ouellet 2008a, likely do not realize that a passage that seems to present a new and original insight from Cardinal Ouellet on a perceived threat faced by the modern media is really a pastiche composed of the words of an archbishop, the current pope, and an unnamed writer at the VIS. The roles of Celli, Benedict XVI, and the VIS writer have been completely effaced in the plagiarizing version. Since only a few months passed between the issuance of the VIS brief in January 2008 and its incorporation into Ouellet 2008a, the plagiarizing ghostwriter again did not hesitate to extract content from relatively recent materials.

Table 5.8 Ratzinger 2005 in Ouellet 2008a

[Cardinal Ouellet's Ghostwriter] Ouellet 2008: 94	Cardinal Ratzinger 2000
The Church always evangelizes and has never interrupted the path of evangelization. She celebrates the Eucharistic mystery every day, administers the sacraments, proclaims the word of life – the Word of God, and commits herself to the causes of justice and charity. And this evangelization bears fruit: It gives light and joy, it gives the path of life to many people; many others live, often unknowingly, of the light and the warmth that radiate from this permanent Evangelization. At the beginning of his public life Jesus says: I have come to evangelize the poor (Luke 4:18); this means: I have the response to your fundamental question; I will show you the path of life, the path toward happiness – rather: I am that path. [...] Jesus preached by day, by night he prayed. His entire life was – as demonstrated in a beautiful way in Luke's Gospel – a path toward the cross, a journey up to Jerusalem. Jesus did not redeem the world with beautiful words but with his suffering and his death. His Passion is the inexhaustible source of life for the world; the Passion gives power to his words. [...] New Evangelization means: to dare, once again and with the humility of the small grain, to leave up to God the when and how it will grow (Mark 4:26-29). The sources are hidden – they are too small. In other words: large realities begin in humility. In the process of the New Evangelization, we are often faced with the great temptation of impatience, the temptation of immediately finding the great success, in finding large numbers. But this is not God's way. For the Kingdom of God as well as for authentic Evangelization, the instrument and vehicle of the Kingdom of God, the parable of the grain of mustard seed is always valid (see Mark 4:31-32). The deepest poverty is the inability of joy, the tediousness of a life considered absurd and contradictory. This poverty is widespread today, in very different forms in the materially rich as well as the poor countries. The inability of joy presupposes and produces the inability to love, produces jealousy, avarice – all defects that devastate the life of individuals and of the world. This is precisely why we are in need of a New Evangelization – if the art of living remains an unknown, nothing else works. But this art is not the object of a science – this art can only be communicated by one who has life – he who is the Gospel personified.	The Church always evangelizes and has never interrupted the path of evangelization. She celebrates the eucharistic mystery every day, administers the sacraments, proclaims the word of life — the Word of God, and commits herself to the causes of justice and charity. And this evangelization bears fruit: It gives light and joy, it gives the path of life to many people; many others live, often unknowingly, of the light and the warmth that radiate from this permanent evangelization. [...] At the beginning of his public life Jesus says: I have come to evangelize the poor (Luke 4:18); this means: I have the response to your fundamental question; I will show you the path of life, the path toward happiness — rather: I am that path. [...] Jesus preached by day, by night he prayed — this is not all. His entire life was — as demonstrated in a beautiful way by the Gospel according to St. Luke — a path toward the cross, ascension toward Jerusalem. Jesus did not redeem the world with beautiful words but with his suffering and his death. His Passion is the inexhaustible source of life for the world; the Passion gives power to his words. [...] New evangelization means: [...] to dare, once again and with the humility of the small grain, to leave up to God the when and how it will grow (Mark 4:26-29). [...] The sources are hidden — they are too small. In other words: The large realities begin in humility. [...] Yet another temptation lies hidden beneath this — the temptation of impatience, the temptation of immediately finding the great success, in finding large numbers. But this is not God's way. For the Kingdom of God as well as for evangelization, the instrument and vehicle of the Kingdom of God, the parable of the grain of mustard seed is always valid (see Mark 4:31-32). [...] The deepest poverty is the inability of joy, the tediousness of a life considered absurd and contradictory. This poverty is widespread today, in very different forms in the materially rich as well as the poor countries. The inability of joy presupposes and produces the inability to love, produces jealousy, avarice — all defects that devastate the life of individuals and of the world. This is why we are in need of a new evangelization — if the art of living remains an unknown, nothing else works. But this art is not the object of a science — this art can only be communicated by [one] who has life — he who is the Gospel personified.

Table 5.9 A VIS brief and Ouellet 2008a

Ouellet 2008: 96-97	VIS 2008
You must influence your brothers and sisters and colleagues who work in the secular media and help them to avoid the risk of being transformed into systems aimed at subjecting humanity to agendas dictated by the dominant interests of the day. This is the challenge facing the media, the challenge we must all face in our daily lives in order to become men and women who show solidarity to all mankind.	The president of the Pontifical Council for Social Communications [Archbishop Claudio Maria Celli] dwelt on "the Pope's clear awareness and knowledge of the fact that unfortunately the media 'risk being transformed into systems aimed at subjecting humanity to agendas dictated by the dominant interests of the day'. This is the challenge facing the media, the challenge we must all face in our daily lives in order to become men and women who show solidarity to all mankind".

The undisclosed sources that constitute Ouellet 2008a are not all magisterial texts, however. Major uncredited sources include a December 2007 article by journalist Sandro Magister (that analyzes some developments in the church's approach to evangelization) and a January 2008 article from the Catholic news agency Zenit (that reported on Benedict XVI's message for World Communications Day) (Ouellet 2008a: 95, column c, lines 43–96 to 96, column a, line 7; 96, column a, lines 18–49 = Magister 2007; Ouellet 2008a: 96, column c, lines 7–42 = Zenit 2008).

5.3 A Second Production by Ouellet's Plagiarizing Ghostwriter

In these cases, texts from the work of journalists reappear in Ouellet's article and thereby acquire a new ecclesiastical endorsement. This "upgrade" by being incorporated into a magisterial text changes the status of the texts by bestowing upon them an authoritative quality they previously lacked. Just as Ouellet 2007 upgraded passages by professional theologians (e.g., Schillebeeckx, Kowalewski) and bestowed upon their words a magisterial authority by being included in the cardinal's address, Ouellet 2008a so upgrades the passages by various journalists (e.g., Magister).

5.4 A Third Production by Ouellet's Plagiarizing Ghostwriter

Although this chapter cannot provide a comprehensive analysis of the literary output of Cardinal Ouellet's ghostwriter, a disclosure of a third instance of plagiarism by the ghostwriter suffices to complete the analysis of the pernicious effects of this particular form of magisterial plagiarism. In this third work, Ouellet's ghostwriter has not exhibited the same degree of plagiarizing proficiency as is found in the two prior compilations. The manufactured text is not as polished as Ouellet 2007 and Ouellet 2008a, and it even offers public clues that it is fraudulent.

The third plagiarizing construction by Ouellet's ghostwriter is a homily given for a mass celebrated at the same International Catholic Media Convention that occasioned Ouellet 2008a. This 2008 homily, the text of which was published online shortly after it was presented, is a sloppy amalgam of both magisterial and non-magisterial source texts (Ouellet 2008b, c). As a homily byOuellet, the text is undoubtedly an exercise of the cardinal's magisterial teaching authority. Insofar as the text of the homily has been published online in two venues, it exercises an influence far beyond the initial audience at the mass.

Ouellet's homily begins with an architectural description of St. Paul's Basilica, a Roman Catholic place of worship in downtown Toronto, where the homily was delivered. The text moves quickly from a discussion of the physical constitution of the building to spiritual matters. Table 5.10 presents the passage where this transition is

Table 5.10 Writings by John Paul II in Ouellet 2008b

Ouellet 2008b	John Paul II 1999; 1998
The significance of the material building lies in the fact that it speaks to us of that superior reality which is "God's building" (I Cor 3:9) "made of living stones" (cf. I Pt 2:5). Here the holy liturgy is celebrated, in which the pilgrim Church on earth expresses the spiritual bond which unites her with the Church in heaven through the communion of saints. On the basis of Baptism, the First Letter of Peter urges Christians to gather round Christ to help build the spiritual edifice founded by and on him: "Come to him [Christ], to that living stone, rejected by men but in God's sight chosen and precious; and like living stones be yourselves built into a spiritual house, to be a holy priesthood, to offer spiritual sacrifices acceptable to God through Jesus Christ" (2:4-5).	The significance of the material building lies in the fact that it speaks to us of that superior reality which is "God's building" (1 Cor 3:9) "made of living stones" (cf. 1 Pt 2:5). Here the holy liturgy is celebrated, in which the pilgrim Church on earth expresses the spiritual bond which unites her with the Church in heaven through the communion of saints. // On the basis of Baptism, the First Letter of Peter urges Christians to gather round Christ to help build the spiritual edifice founded by and on him: "Come to him [Christ], to that living stone, rejected by men but in God's sight chosen and precious; and like living stones be yourselves built into a spiritual house, to be a holy priesthood, to offer spiritual sacrifices acceptable to God through Jesus Christ" (2:4-5).

made, and the table discloses the two uncredited source texts that are being plagiarized in the homily.

The transition in 2008b from a discussion of the physical building of St. Paul's Basilica to the spiritual edifice of the church includes Ouellet's line "here the holy liturgy is celebrated." The expression *holy liturgy* in this context is odd. St. Paul's Cathedral in Toronto is a Roman Catholic church that offers masses celebrated according to the Roman rite. The expression "holy liturgy" typically refers to celebrations of the Eucharist offered according to the Byzantine Rite in both Orthodox and Byzantine Catholic churches. This inappropriate idiom by Ouellet's ghostwriter indicates that something is amiss. Ouellet's ghostwriter has plagiarized verbatim from John Paul II's remarks given on the occasion of the pope's 1999 visit with Ilia II, the Patriarch of the Georgian Orthodox Church (John Paul II 1999). The visit occurred in Svetitskhoveli Cathedral in Mtskheta, northwest of the Georgian capital Tbilisi. The building that John Paul II is discussing in his public remarks is the patriarchal cathedral, where of course the "holy liturgy" is celebrated according to the Byzantine rite. In taking the pope's remarks, the plagiarizing ghostwriter retained the language proper to that specific context and occasion, thereby producing an infelicitous plagiarizing text.

That dense patchwork style of Ouellet's ghostwriter is emblematically exhibited in Table 5.11. The lengthy passage is a blend of sentences from three unacknowledged source works. The main source is a 2003 essay by Lutheran pastor Mary W. Anderson that was published in the mainline Protestant magazine *The Christian Century*. The two other sources are a homily by Franciscan priest Raniero Cantalamessa, who holds the curial office of the Preacher of the Papal Household, and a homily by Benedict XVI from his 2008 visit to St. Patrick's Cathedral in New York.

The first part of Ouellet's text is a result of the ghostwriter's refashioning of the Pope's remarks to fit the new occasion of the address to those who work in communications and the media. Notably, there are cited quotations from Benedict XVI's homily earlier in Ouellet 2008b, but the portion exhibited in Table 5.11 is not credited to the pope. More interesting is the way the non-magisterial sources are employed in the plagiarizing passage. Although Cantalamessa does not himself enjoy episcopal consecration and therefore does not possess magisterial teaching authority, he does in virtue of his office as *Praedicator sacri palatii* have the unique ecclesiastical position of preaching to the pope and members of the Roman Curia. By plagiarizing Cantalamessa's homily, Ouellet's ghostwriter has taken the specific words that were written to address Benedict XVI and other members of the Roman Curia.

Almost 18% of Ouellet's address consists of verbatim and near-verbatim extracts from the essay by pastor and Lutheran theologian Mary W. Anderson. The title of her essay in *The Christian Century* ("Blind Spots") does not reveal that it is a critical appreciation of the Protestant Reformation, which she views in light of the lectionary reading from Mark's gospel about the healing of the blind beggar Bartimaeus. Ouellet's ghostwriter has discarded the portions of the article dealing with the Reformation and has only retained the scriptural exegesis. That the words of a Lutheran minister are forming much of the basis of the cardinal's homily may come

5.4 A Third Production by Ouellet's Plagiarizing Ghostwriter

Table 5.11 The patchwork plagiarism style of Cardinal Ouellet's ghostwriter

Ouellet 2008b	Benedict XVI 2008; Anderson 2003: 20; Cantalamessa 2006; Anderson 2003; Cantalamessa 2006
You have no easy task in being excellent journalists and communicators when you work in a world that often looks at the Church "from the outside", - a world which deeply senses a need for spirituality, yet finds it difficult to "enter into" the mystery of the Church. Even for those of us within, and those who "cover the Church on a daily basis," the light of faith can be dimmed by routine, and the splendor of the Church obscured by the sins and weaknesses of her members. This leads me to my second thought that I wish to share with you on this evening's moving Gospel story from St. Mark- the healing of Bartimaeus the blind man. Healing stories in the Gospels never seem to be simply a reversal of physical misfortune. A paralyzed man stands and walks. A man stretches out a withered hand to Jesus and sees it become useful again. A girl who was pronounced dead awakens. Particularly suspicious are the stories of those who "once were blind, but now they see." The connections between seeing and believing are so strong in the Gospel accounts that these miracles worked through Jesus almost always seem more about growing in faith than taking off dark glasses. Though Bartimaeus was blind to many things, he clearly saw who Jesus was. But Bartimaeus is not blind; he is only sightless. He sees better with his heart than many of those around him, because he has faith and cherishes hope. More than that, it is this interior vision of faith that also helps him to recover his external vision of things. "Your faith has made you well," Jesus says to him. Seeing "who Jesus is" is the goal of faith, and it leads to discipleship. At the end of the story we're told that this is exactly what happened. Bartimaeus regained his sight and followed Jesus on the way. Given that the very next verse in Mark narrates the entry into Jerusalem, the way Bartimaeus followed was the way to the cross. Bartimaeus, like every good communicator and journalist, does not miss an opportunity! He heard that Jesus was passing by, understood that it was the opportunity of his life and acted swiftly.	This is no easy task in a world which can tend to look at the Church, like those stained glass windows, "from the outside": a world which deeply senses a need for spirituality, yet finds it difficult to "enter into" the mystery of the Church. Even for those of us within, the light of faith can be dimmed by routine, and the splendor of the Church obscured by the sins and weaknesses of her members. // Healing stories in the Gospels never seem to be simply a reversal of physical misfortune. A paralyzed man stands and walks. A man stretches out a withered hand to Jesus and sees it become useful again. A girl who was pronounced dead awakens. Particularly suspicious are the stories of those who "once were blind, but now they see." The connections between seeing and believing are so strong in the Gospel accounts that these miracles worked through Jesus almost always seem more about growing in faith than taking off dark glasses. Though Bartimaeus was blind to many things, he clearly saw who Jesus was. // Bartimaeus is not blind; he is only sightless. He sees better with his heart than many of those around him, because he has faith and cherishes hope. More than that, it is this interior vision of faith which also helps him to recover his external vision of things. "Your faith has made you well," Jesus says to him. // Seeing "who Jesus is" is the goal of faith, and it leads to discipleship. Only the unblind can see where to follow. Indeed, at the end of the story we're told that this is exactly what happened. Bartimaeus regained his sight and followed Jesus on the way. Given that the very next verse in Mark narrates the entry into Jerusalem, the way Bartimaeus followed was the way to the cross. // Bartimaeus is someone who does not miss an opportunity. He heard that Jesus was passing by, understood that it was the opportunity of his life and acted swiftly.

as a surprise to many. Anderson's words are enjoying an afterlife in the context of a Catholic mass that likely neither she nor her early readers would have anticipated.

Due to the ghostwriter's work, Ouellet presents in the homily some first-person claims that were originally expressed by Anderson in the first person. The "I" of the genuine author Anderson, described in *The Christian Century* with the byline "Mary W. Anderson is the pastor of Incarnation Lutheran Church in Columbia, South Carolina," is not the "I" who is the Cardinal of Quebec and Primate of Canada. Certainly both possess an authority for those under their spiritual care. Yet the authority of the latter, conceived in light of the Catholic church's teaching of magisterial authority, does not attach to Anderson's words as expressed in the article in *The Christian Century*.

5.5 The Downgrading of Magisterial Texts

An opposite phenomenon occurs when a plagiarizing theologian who lacks magisterial authority takes a magisterial text and presents it as if it were a new work in theology. In such cases, the magisterial text is presented without its magisterial endorsement. Because its magisterial quality is concealed, traditional Catholic readers who encounter the plagiarizing work may regard it as merely a work in theology, rather than as an authoritative text requiring some degree of adherence. Through this kind of plagiarism, magisterial texts are *downgraded* by being presented as mere works of theology rather than as magisterial texts.

An extensive case of serial plagiarism exemplifies this phenomenon of magisterial downgrading. In 2019, many (if not most) of the extensive academic and journalistic works of a priest (hereafter, "R.") were revealed to be plagiarizing compilations of the works of others. Editors and publishers issued more than 30 corrections—including 23 retractions—that earned R. the ignominy of a listing on *The Retraction Watch Leaderboard* (Retraction Watch 2019; Oransky 2019). R.'s plagiarizing articles appeared in a range of venues, from academic theological journals to more popular outlets for writings on religious themes. For example, *Origins*—the same publication that issued the three plagiarizing articles by Cardinals Levada and Ouellet—published eight of the plagiarizing articles by R. in the period of 2009–2017. To each of R.'s articles in *Origins*, the editors appended a brief notice stating that each one "contains numerous instances of the words of others without proper citation," and these short notifications were supplemented by a comprehensive statement covering all eight that was published in mid-2019 (Editors of *Origins* 2019: 176). In light of these 11 defective articles published under the names of "R.," "Levada," and "Ouellet," one is tempted to conclude that *Origins* has been established as significant gateway for the proliferation of plagiarizing content in the discipline of Catholic theology in recent years.

In producing his plagiarizing articles for over a 30-year period, R. appropriated texts from a variety of sources, including many magisterial documents. In appropriating the words of popes, cardinals, and bishops as his own throughout his writings, R. presented many passages from magisterial texts without the papal or episcopal attestation that originally attended their promulgation. This form of magisterial plagiarism puts readers at a disadvantage in assessing the meaning and weight of the plagiarizing texts (see Dougherty and Hochschild 2019, 2020). On a traditional Catholic view, works written by theologians, insofar as they are theologians, do not possess a magisterial quality. Those works, however, that are issued by members of the Roman Catholic hierarchy in the exercise of their teaching function do possess such a quality. Catholic readers of R.'s plagiarizing compilations are disadvantaged in two ways: they are deceived about the source of the words, and they are denied the opportunity to interpret them in light of their original status as magisterial.

Dougherty and Hochschild 2020 have provided a detailed account of many examples where R. presents passages from magisterial texts as if they were his own writings. R.'s 2012 biblical commentary *Where Jesus Walked* is a representative example of R.'s plagiarizing compilations. Published by the Canadian Conference of Catholic Bishops (CCCB), the book consists of expositions of passages of the gospels following the order of the Roman Missal. Many of the reflections in the book that appear to be original by R. are really plagiarized from dozens of sources, many of which are magisterial. After being notified of the plagiarism by third-party whistleblowers, the Canadian Conference of Catholic Bishops issued a retraction notice for the book and two others in June 2019 that stated, "Rev. [R.] [...] failed to provide all the appropriate citations, as well as bibliographic references, and did not acknowledge a number of original sources" (CCCB 2019). The retraction statement did not identify any of the individual authors whose works were appropriated in the plagiarism, so no indication is given that many of R.'s undocumented sources are passages extracted from the magisterial writings of popes, cardinals, and bishops.[2]

5.6 Who Is the Plagiarizing Ghostwriter for Cardinal Ouellet?

The identities of ecclesiastical ghostwriters who work for popes, cardinals, and bishops generally remain concealed. One unusual exception was the disclosure of Archbishop Victor Manuel Fernández as a ghostwriter for Pope Francis's post-synodal apostolic exhortation *Amoris Laetitia*. This outing occurred because of textual evidence present in the apostolic exhortation itself. Had Fernández not placed portions of his own earlier theological writings into the document, he likely would never have been brought into the spotlight, thereby avoiding subsequent criticism. The textual evidence for Fernández's ghostwriting role was further supplemented by loose circumstantial evidence that commentators observed; Francis and Fernández have enjoyed a longstanding relationship going back decades. The combined strength of the textual and circumstantial evidence led journalist Sandro Magister and philosopher Michael Pakaluk to each publish findings that Fernández was a ghostwriter for *Amoris Laetitia*, and Fernández's role in the construction of other papal documents by Francis is now generally recognized (Magister 2016; Pakaluk 2017).

Considering this precedent, one might wonder whether the unnamed ecclesiastical ghostwriters for the compilations produced for Cardinals Levada and Ouellet examined above could be similarly identified, based on latent textual evidence in the published documents in addition to any circumstantial evidence. Since at present only one of Levada's articles has been shown to be plagiarized, whereas forOuellet there are at least three, the known field of potential evidence for the identity of Ouellet's plagiarizing ghostwriter is larger. This section considers the circumstantial and textual evidence for the identity of Cardinal Ouellet's ghostwriter for Ouellet 2007, 2008a, b, proposing that that R. is likely a ghostwriter for those articles.

---

[2]A further variation of theological plagiarism occurs when a prelate has committed plagiarism in a publication prior to having acquiring episcopal ordination. For this possibility, see Schachenmayr 2019.

5.6.1 The Circumstantial Evidence

There are loose pieces of circumstantial evidence suggesting the possibility that R. and Ouellet's ghostwriter are one and the same person. First, both have been revealed as serial plagiarists who specialize in producing works of Catholic theology. Presumably the pool of accomplished serial plagiarists in Catholic theology who are contemporaneous is not exceptionally large. Second, both R. and Ouellet's ghostwriter use the same patchwork-style plagiarism. The methodologies of the two are the same, yet there are many ways to plagiarize (see Weber-Wulff 2014: 6–12; Gipp 2014: 11–13, and Chap. 1). Third, Ouellet and R. know each other, apparently to the extent that they are not mere acquaintances. R. has expressed in various fora that he has known Ouellet for many years and that he "worked closely" with the cardinal on several projects (R. 2010). In 2012, R. interviewed Ouellet, and in formulating a question R. revealed, "I know you. I've known you for a long time. I consider you a friend, a brother priest, and, and a great leader" (R. 2012c: 22:01). Fourth, within the first two paragraphs of both Ouellet 2007 and 2008a, the cardinal mentions R. by name (Ouellet 2007: 60, 2008a: 93). To find stronger evidence, however, for the identity of R. and Ouellet's ghostwriter, one must turn to the plagiarizing texts themselves.

5.6.2 The Textual Evidence

In the April 23, 2004 installment of his weekly "Word from Rome" column for *National Catholic Reporter*, Vatican journalist John L. Allen, Jr. included an account of a lecture by Passionist priest Donald Senior, the then-president of the Catholic Theological Union in Chicago and member of the Pontifical Biblical Commission. Allen summarized points of Senior's lecture and included a few short quotes from Senior. Allen's column is one of the uncredited source texts for Ouellet 2007, which appropriates both Allen's analysis and Senior's quotes, presenting them as new reflections. Table 5.12 provides an example.

Apart from the verbatim plagiarism of Allen's summary of Senior's lecture, what may be most surprising is the transformation of the quotations by Senior. They have been reworked to give the illusion of a seamless commentary, having been stripped of quotation marks and any identifying reference to their speaker. One who reads

Table 5.12 Allen 2004 and Ouellet 2007

[Cardinal Ouellet's Ghostwriter] Ouellet 2007: 60	Allen 2004
Terrorism, along with ethnic and religious divisions, generates violence that seems to have no end. Economic insecurity raises collective anxieties. [...] We need to recover the depth, beauty and vastness of the church's mission. This is not a time for hesitation or retreat. We need to keep the arena large [...].	Terrorism, along with ethnic and religious divisions, generates violence that seems to have no end. Economic insecurity raises collective anxieties. The church is suffering as well. [...] Against this backdrop, [Fr. Donald] Senior suggested, Christians need to recover the "depth and beauty" of its mission. "This is not a time for hesitation or retreat," Senior said. "We need to keep the arena large."

5.6 Who Is the Plagiarizing Ghostwriter for Cardinal Ouellet?

in Ouellet 2007 "This is not a time for hesitation or retreat" and "We need to keep the arena large" will not know that these were sentences presented by Senior in his 2007 lecture that were then extracted with full citations in Allen's subsequent column about the lecture. This context is made to disappear due to the plagiarizing work of Ouellet's ghostwriter, resulting in the words of Allen and Senior appearing in Ouellet 2007 as if they were original insights by the cardinal.

Such transformations of Allen's column are highly idiosyncratic. Nevertheless, these same transformations appear first in a plagiarizing lecture by R. in late 2006, prior to the 2007 plagiarizing address by Ouellet (R. 2006). That 2006 address by R. postdates Allen's 2004 column yet precedes Cardinal Ouellet's 2007 address by just over two and a half months. What is more, R. continued to re-use a plagiarizing idiosyncratic reformation of Allen's column in later publications for at least six more years (e.g., R. 2008: 227, 2012b: 13). The presence of the idiosyncratic plagiarizing reformulation of Allen's article first in the plagiarizing text of R. 2006, then in Ouellet 2007, followed again in later various plagiarizing articles by R., can be interpreted in two ways. Either R. is Ouellet's ghostwriter, or Ouellet's ghostwriter is plagiarizing R.

This evidence appears in a new light if one considers that there are *many* passages common to works by Ouellet's ghostwriter and works by R. The two sets of contemporaneous works—the corpus of Ouellet's ghostwriter and the corpus of R.—plagiarize from a relatively small pool of similar sources. For example, both R.'s 2012 book *Where Jesus Walked* and Ouellet 2008b each draw copiously from the obscure article in *The Christian Century* by Lutheran pastor and theologian Mary W. Anderson discussed earlier (Anderson 2003). Anderson's words appear without attribution on page 53 of R.'s *Where Jesus Walked*, interwoven with uncredited sentences taken from a homily by Cantalamessa. Yet R. is not copying Ouellet 2008b, as the book plagiarizes some portions of Anderson's article not found in Ouellet 2008b. Furthermore, the sentences from Cantalamessa interspersed with Anderson's sentences in R.'s 2012 book are from a different homily by the Preacher of the Papal Household. R. and Ouellet's ghostwriter are plagiarizing from the same set of sources here and not from each other.

Another common source from which both R. and Ouellet's ghostwriter plagiarize is Archbishop Rylko's 2006 address on ecclesial movements and communities. Four paragraphs of Rylko's address are found in compilations by each plagiarist (Ouellet 2008a; R. 2011: 374, R. 2009: 496–497, 499). As far as magisterial texts go, Rylko's 2006 address is arguably somewhat obscure, yet both plagiarists favor it in their plagiarizing compilations.

Often the same sentences inexplicably appear in works by R. and Ouellet's ghostwriter. Sometimes the directionality of the migration of the idiosyncratic sentences is from Ouellet's ghostwriter to R. For example, the very particular formulation "to re-actualize the historic and cultural patrimony of holiness and social engagement of the Church which draws its roots from the Eucharistic mystery" shows up first in Ouellet 2007: 60 but then later in R. 2010. But sometimes sentences migrate in the opposite direction, from an initial appearance in a work by R. to a later writing by Ouellet's ghostwriter. For example, the description of John Paul II's 2002 address

Table 5.13 R. 2006 and Ouellet 2007 on John Paul II

R. 2006	Ouellet 2007: 60
Pope John Paul II spoke these thought-provoking words to the crowd of 600,000+ young people gathered at the Great Vigil of World Youth Day 2002: "The new millennium opened with two contrasting scenarios […]. Is it right to be content with provisional answers to the ultimate questions, and to abandon life to the impulses of instinct, to short-lived sensations or passing fads?" There could not be more fitting images to describe the awesome backdrop of our work of evangelization in the Church today. Our contemporaries are living in a world that is suffering from tremendous pain and loss. The striking images evoked by the Pope remain engraved on people's memories.	Pope John Paul II spoke these thought-provoking words to the crowd of 600,000-plus young people gathered at the great vigil of World Youth Day 2002: "The new millennium opened with two contrasting scenarios […]. Is it right to be content with provisional answers to the ultimate questions and to abandon life to the impulses of instinct, to short-lived sensations or passing fads?" There could not be more fitting images to describe the stark backdrop of our work of evangelization in the church today. The striking images evoked by the pope remain engraved on people's memories.

at World Youth Day in Toronto that is found in a December 2006 work by R. also appears also following April in a work by Ouellet. Table 5.13 displays the passage common to both.

These two texts overlap and were issued only months apart. And yet two months later, after appearing first in R. 2006, and then in Ouellet 2007, a portion of the same passage reappears, but now in a July interview given by R. (2007). The extremely short timeline is relevant; within 7 months, what is fundamentally the same passage has appeared in a work by R., then in a work by Ouellet's ghostwriter, and then again in another work by R.

Perhaps strongest textual evidence for the identity of R. and Ouellet's ghostwriter is that R. plagiarizes heavily from Ouellet 2007 in a biblical commentary on the lectionary he published under his own name in 2011. Since Ouellet 2007 is itself a plagiarism of the texts of various sources, those same texts of various sources therefore appear also in the 2011 biblical commentary (R. 2011b). Table 5.14 not only exhibits overlap between Ouellet 2007 and R. 2011b, but further illustrates that source passages from theologians Jeremy Driscoll, Igor Kowalewski, and Edward Schillebeeckx that were plagiarized in Ouellet 2007 also appear in R. 2011b.

There are no words of Driscoll, Kowalewski, or Schillebeeckx in R. 2011b that are not also in Ouellet 2007. Ouellet 2007 must be the uncredited source for R. 2011b.

5.6.3 Triple Plagiarism

This disclosure that a portion of Ouellet 2007 reappears without credit R. 2011b also reveals yet another stage to the exotic journey of the words of theologian Jeremy Driscoll. Having appeared in two plagiarizing compilations without credit (Levada 2007; Ouellet 2007), the words are now exhibited to have appeared in a third: R. 2011b. In their additional appearance there, Driscoll's words are shorn of the magisterial attestation that they had temporarily gained when they were included in the plagiarizing compilations issued by Cardinals Levada and Ouellet. Now Driscoll's words show up in a further plagiarizing iteration, more than a decade after their original publication, but this time by one who lacks episcopal authority. In this final display, the words no longer enjoy a magisterial status; they are relegated back to

5.6 Who Is the Plagiarizing Ghostwriter for Cardinal Ouellet?

Table 5.14 Ouellet 2007 as an undisclosed source text for R. 2011b

R. 2011b	Ouellet 2007: 62, 63, 62
The most important doctrines of our Catholic Christian faith remain the same through the ages and need to be approached again and again in order to rediscover their richness and experience their enduring significance for our daily lives. These doctrines are the deepest sense of what the Scriptures proclaim and that this deepest sense is discovered precisely when the Scriptures are proclaimed in the liturgical assembly and when the Scriptures become sacrament in the Eucharistic rite. From this source we draw our energy, our vision and our hope to foster a true civilization of love. At every mass, the liturgy of the Word precedes the Eucharistic liturgy. There are two "communions," one with the Word and one with the Bread. One cannot be understood without the other. The Eucharist does not only provide inner strength, but also a certain way of life. It is a way of living that is passed from Jesus to the Christian. The celebration of the Eucharist has no meaning if it is not lived with love. Through the Eucharist we are challenged at the level of our history to realize as much as possible what we celebrate sacramentally: bread for all, salvation and liberation for all. The Eucharistic Christ is truly present as bread for the poor, and not for the privileged. In order to keep the Eucharistic reality credible, we have to devote ourselves to a better, more just world.	The most important doctrines remain the same through the ages and need to be approached again and again by theology and in our preaching; namely, the divine and human natures of Christ; their union in the divine person of the Son; and the mystery of the Holy Trinity that Christ reveals in his paschal mystery. These doctrines are the deepest sense of what the Scriptures proclaim and that this deepest sense was discovered precisely when the Scriptures were proclaimed in the liturgical assembly and when the Scriptures became sacrament in the eucharistic rite. From this source we draw our energy, our vision and our hope to foster a culture of the Eucharist and a civilization of love. [...] At every Mass, the Liturgy of the Word of God precedes the eucharistic liturgy. There are two "communions," one with the word and one with the bread. One cannot be understood without the other. [...] The Eucharist does not only provide inner strength but also a certain way of life. It is in fact a way of living that is passed from Jesus to the Christian. [...] Through the Eucharist we are challenged at the level of our history to realize as much as possible what we celebrate sacramentally: bread for all, salvation and liberation for all, including greater respect for creation and promotion of a safer environment. The eucharistic Christ is really present as bread for the poor. The Christ who is sacramentally present is a constant reminder and pledge that our sharing of the bread is not in vain. In order to keep the eucharistic reality credible, we have to devote ourselves to a better, more just and habitable world.
Highlighting = Overlap with between R. 2011b and Ouellet 2007: 62, 63 Bold underlining = Overlap with Driscoll 2000: 32; 2002: 31–32; 2003: 224–225 Dotted underlining = Overlap with Kowalewski 2005 Wavy underlining = Overlap with Schillebeeckx 1990: 78	

their normal theological status. That is, words have returned to the non-magisterial state under which they were first issued, but without correct attribution to Driscoll.

With this loss of magisterial attestation to Driscoll's words, things have come full circle. Figure 5.3 illustrates the progression of the plagiarism of Driscoll's words through the five plagiarizing iterations.

Only stages 2 and 4, namely, the published works Levada 2007 and Ouellet 2007, can be considered magisterial. By stage 5, Driscoll's words have lost the magisterial quality they had possessed in the earlier published plagiarizing iterations. This extensive journey of Driscoll's words is less astonishing if Ouellet's ghostwriter and R. are the same person, since a single individual would be responsible for most of the stages (3, 4, and 5).

The respective published works of R. and Ouellet's ghostwriter are overlapping in idiosyncratic ways, both in terms of common sentences and paragraphs, but also in terms of source texts for the acts of plagiarism. There are two possibilities. Either R. and Ouellet's ghostwriter are the same person or they not. To assume the *Non-Identity Thesis* entails that there are two contemporaneous serial plagiarists in Catholic theology who have an inexplicable devotion to the same very narrow subset of source texts for their plagiarism, covering both magisterial sources and non-magisterial ones, some of which are quite obscure. The two contemporaneous serial plagiarists both utilize the same patch-work style plagiarism, and they regularly plagiarize not only from the same source texts but also from each other's plagiarizing compilations. Subscribers to this *Non-Identity Thesis* must view all instances of identical

Fig. 5.3 Driscoll's text in five plagiarizing compilations

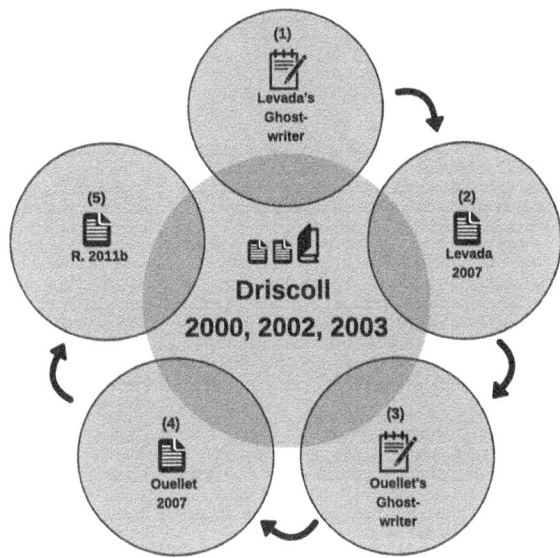

idiosyncratic manipulations of source texts in the plagiarizing works of both R. and Ouellet's ghostwriter to be the result of repeated instances where one plagiarist is plagiarizing the work of the other.

In contrast, to assume the *Identity Thesis* simply entails that the defective plagiarizing works are the productions of one active serial plagiarist in Catholic theology who uses a consistent style of patchwork plagiarism. The apparent devotion to same very narrow subset of source texts in the plagiarizing works of R. and Ouellet's ghostwriter is only because R. publishes some of his compilations under his own name and offers some of his other compilations to prelates.

It is not possible at present to establish with demonstrable certainty that Cardinal Ouellet's ghostwriter is R. Unless the plagiarizing ghostwriter for Ouellet steps forward, or new evidence is uncovered, the evidence is not dispositive. Nevertheless, the principle of parsimony favors the *Identity Thesis* over the *Non-Identity Thesis*. Considering the evidence known thus far, it is simpler to assume that the ghostwriter for Cardinal Ouellet is none other than R.

5.7 Conclusion

The use of ghostwriters by members of the Catholic hierarchy in producing magisterial documents is neither controversial nor unusual. Popes, cardinals, and bishops regularly employ them, just as politicians employ speechwriters or judges employ law clerks who help draft legal decisions. In these three distinct contexts—ecclesiastical, political, judicial—there is no expectation that a finalized document is solely the production of a given bishop, politician, or judge. Since magisterial texts are promulgated by the authority of popes, cardinals, or bishops, the employers of the ghostwriters rather than the ghostwriters themselves appear as the authors of record for the magisterial documents, much like politicians are treated as the authors of their speeches, and judges are treated as the authors of their legal opinions. In other contexts, such as educational settings, ghostwriting is judged to be a severe violation of research integrity. There has been an increased scrutiny of domains—such as in medical and pharmaceutical publishing—where ghostwriting is considered to be an especially pernicious practice that corrupts the reliability of published findings and the subsequent medical regimens based upon them (Crombie 2020).

The use of ghostwriters in the production of magisterial documents becomes controversial, however, only when ghostwriters commit a violation of the expected norms of professional ghostwriting. Plagiarism is one grave kind, but there are others. As noted above, when Pope Francis released *Amoris Laetitia*, some observers pointed out that the document included sentences from several theology articles by Fernández. As one critic put it, Fernández failed his papal employer because "a ghostwriter should remain a ghost" (Pakaluk 2017).

This chapter has identified the two principal forms of magisterial plagiarism. First, when ghostwriters produce plagiarizing documents for their ecclesiastical employers, the subsequent promulgation of them by a pope, cardinal, or bishop upgrades the various source texts by bestowing on them a new magisterial attestation. The original source texts thereby receive a non-negligible upgrade in their relevance for the practice of Catholic theology. The presence of plagiarizing material in magisterial documents is not inconsistent with the view in the Catholic tradition that the content of magisterial documents is presumed to be true, reliable, and authoritative. The CDF has been clear that "it could happen that some Magisterial documents might not be free from all deficiencies" (Congregation for the Doctrine of the Faith 1990: §24). Second, when plagiarizing theologians pass off portions of previously issued magisterial documents as their own, the magisterial documents receive a downgrade insofar as their magisterial attestation is concealed to readers. Those seeking to practice theology in the manner of *sentire cum ecclesia* are disadvantaged in assessing the weight of these texts in the specific context of Catholic theology.

References

Allen, Jr., John L. 2004. The word from Rome. *NCR*, April 23. http://www.nationalcatholicreporter.org/word/pfw042304.htm.

Anderson, Mary W. 2003. Blind spots. *The Christian Century 120* (21): 20.

Benedict XVI. 2008. Votive Mass For the universal church. *Congregation for the Clergy*. http://www.clerus.org/bibliaclerusonline/en/cwf.htm#zb.

Benedict XVI. 2011. General Audience. 14 December. *Vatican.va*. http://w2.vatican.va/content/benedict-xvi/en/audiences/2011/documents/hf_ben-xvi_aud_20111214.html.

Cantalamessa, Raniero. 2006. Chosen from and for men. *Zenit*, October 27. https://zenit.org/articles/father-cantalamessa-on-the-priesthood.

Catechism of the Catholic Church. 1995. New York: Doubleday.
CCCB. 2019. Statement of Retraction. https://www.cccb.ca/site/eng/media-room/announcements/5127-notice-of-retraction.
Congregation for the Doctrine of the Faith. 1990. *Donum veritatis*. *Vatican.va*. http://www.vatican.va/roman_curia/congregations/cfaith/documents/rc_con_cfaith_doc_19900524_theologian-vocation_en.html.
Crombie, James. 2020. Medical ghost- and guest-writing as corrupt practices and how to prevent them. In *Integrity, transparency and corruption in healthcare & research on health*, vol. I, ed. Kıymet Tunca Çalıyurt, 141–158. Singapore: Springer.
Dougherty, M. V., and Joshua Hochschild. 2019. Tracking Fr. [R.]'s (very) long history of plagiarism. *National Post*, April 15. https://nationalpost.com/opinion/tracking-father-rosicas-very-long-history-of-plagiarism.
Dougherty, M. V., and Joshua Hochschild. 2020. Magisterial authority and theological authorship: The harm of plagiarism in the practice of theology. Unpublished manuscript.
Driscoll, Jeremy. 2000. The Fathers and Eucharistic preaching. *Antiphon* 5 (3): 29–38. [Corrigendum in Driscoll 2019].
Driscoll, Jeremy. 2002. Preaching in the context of the Eucharist. *Pro Ecclesia* 11: 24–40.
Driscoll, Jeremy. 2003. *Theology at the Eucharistic table*. Rome: Centro Studi S. Anselmo.
Driscoll, Jeremy. 2019. Corrigendum. https://journals.sagepub.com/doi/full/10.1177/1063851219879524.
Dulles, Avery. 2007. *Magisterium*. Naples, FL: Sapientia Press.
Editors of *Origins*. 2019. On file. Readers. *Origins* 49 (11): 176.
Gipp, Bela. 2014. *Citation-based plagiarism detection*. Wiesbaden: Springer Vieweg.
Harrison, Matthew. 2008. Marc Cardinal Ouellet: The new evangelization and the mass media. *Salt and Light Media*, May 30. http://saltandlighttv.org/blogfeed/getpost.php?id=990.
John Paul II. 1998. General Audience: Wednesday 15 April. *Vatican.va*. https://w2.vatican.va/content/john-paul-ii/en/audiences/1998/documents/hf_jp-ii_aud_15041998.html.
John Paul II. 1999. Greeting to the Catholicos-Patriarch and the Holy Synod. *Vatican.va*, November 8. https://w2.vatican.va/content/john-paul-ii/en/speeches/1999/november/documents/hf_jp-ii_spe_19991108_tbilisi-cathedral.html.
Kowalewski, Igor. 2005. The spiritual development of Christians. *Congregation for the Clergy*. http://www.clerus.org/clerus/dati/2005-02/26-13/03EUING.html.
Levada, William. 2007. The liturgical foundations of theology. *Origins* 36 (42): 687–689.
Levada, William. 2008a. The homilist. *Origins* 37 (38): 601–608.
Levada, William. 2008b. The homilist. *Vatican.va*, February 12 http://www.vatican.va/roman_curia/congregations/cfaith/documents/rc_con_cfaith_doc_20080212_levada-homilist_en.html.
Lumen Gentium. 1964. *Vatican.va*. http://www.vatican.va/archive/hist_councils/ii_vatican_council/documents/vat-ii_const_19641121_lumen-gentium_en.html.
Magister, Sandro. 2007. Overturned: The church can – and must – evangelize. *Chiesa*, December 17. http://chiesa.espresso.repubblica.it/articolo/182761bdc4.html?eng=y.
Magister, Sandro. 2016. 'Amoris Laetitia' has a ghostwriter. *Chiesa*, May 25. http://chiesa.espresso.repubblica.it/articolo/1351303bdc4.html?eng=y.
Mele, Joseph. 2008. Homiletics at the threshold. Ph.D. Dissertation. Duquesne University. https://dsc.duq.edu/etd/919.
Oransky, Ivan. 2019. Plagiarism prompts retraction of 25-year-old article by prominent priest. *Retraction Watch*, March 4. https://retractionwatch.com/2019/03/04/plagiarism-prompts-retraction-of-25-year-old-article-by-prominent-priest.
Ouellet, Marc. 2007. The 49th International Eucharist Congress. *Origins* 37 (4): 59–63.
Ouellet, Marc. 2008a. The New Evangelization and the Mass Media. *Origins* 38 (6): 93–98.
Ouellet, Marc. 2008b. Homily of Marc Cardinal Ouellet. May 29, 2008. *Salt and Light Media*, May 29. https://saltandlighttv.org/blogfeed/getpost.php?id=998.
Ouellet, Marc. 2008c. Cardinal Ouellet's homily. *The Daily Offices*, June 2008. http://thedailyoffices.blogspot.com/2008/06/cardinal-ouellets-homily-may-29-in.html.

References

Pakaluk, Michael. 2017. Ethicist says ghostwriter's role in 'Amoris' is troubling. *Crux*, January 15. https://cruxnow.com/commentary/2017/01/15/ethicist-says-ghostwriters-role-amoris-troubling.

Palmo, Rocco. 2007. Scenes from a sesquicentennial. *Whispers in the Loggia*, March 23. http://whispersintheloggia.blogspot.com/2007/03/scenes-from-sesquicentennial.html.

Ratzinger, Joseph. 2000. The New Evangelization. *Zenit*. https://zenit.org/articles/cardinal-ratzinger-on-the-new-evangelization.

Ratzinger, Joseph. 2007. *Jesus of Nazareth*. Trans. Adrian J. Walker. New York: Doubleday.

Retraction Watch Leaderboard. 2019. *Retraction Watch*, July 17. https://web.archive.org/web/20190717132824, https://retractionwatch.com/the-retraction-watch-leaderboard.

[R.]. 2006. A new springtime. *Catholic Christian Outreach (CCO) Rise up Conference*, December 30, Quebec City.

[R.]. 2007. Interview with Father [R.]. *Zenit*, July 29. https://www.catholic.org/featured/headline.php?ID=4668.

[R.]. 2008. Glimpses of glory. In *Northern light*, ed. Byron Rempel-Burkholder and Dora Dueck, 226–232. Mississauga: Wiley Canada.

[R.]. 2009. Rise up, Young People of Canada. *Origins* 38 (31): 495–499. [Correction issued in *Origins* 49 (11): 176.].

[R.]. 2010. Au revoir et mille mercis, Cardinal Ouellet! *Salt and Light Media*, July 5. https://web.archive.org/web/20101127200558, http://saltandlighttv.org/blog/?p=14569.

[R.]. 2011a. Young people and the New Evangelization. *Origins* 41 (24): 373–378. [Correction issued in *Origins* 49 (11): 176.].

[R.] 2011b. Sacrament of piety, sign of unity, bond of charity. *Zenit*, June 21. https://zenit.org/articles/sacrament-of-piety-sign-of-unity-bond-of-charity.

[R.]. 2012a. *Where Jesus walked*. Ottawa: CCCB. [Retracted in: CCCB 2019].

[R.]. 2012b. Making time for God. *Origins* 42 (1): 9–15. [Correction issued in Origins 49 (11): 176.].

[R.]. 2012c. Cardinal Marc Ouellet – Witness. *Salt and Light Media*, February 12. https://youtu.be/rda69ZGsxrI.

Rylko, Archbishop Stanislaw. 2006. Ecclesial movements and new communities. *Zenit*. https://zenit.org/articles/on-ecclesial-movements-and-new-communities.

Schachenmayr, Alkuin. 2019. Concerns about Bishop Stephen Robson's Dissertation on Bernard of Clairvaux. *Analecta Cisterciensia* 69: 420–428.

Schillebeeckx, Edward. 1990. *For the sake of the Gospel*. Trans. John Bowden. New York: Crossroad.

Schoof, Ted Mark. 2014. Introduction to the New Edition. In *The Collected Works of Edward Schillebeeckx*: Volume II, xxi–xxiv. London: Bloomsbury.

Sullivan, Francis A. 1983. *Magisterium*. Mahwah, NJ: Paulist Press.

Sullivan, Francis A. 1996. *Creative fidelity*. New York: Paulist Press.

The American College. 2007. The American College Sesquicentennial: 1857–2007. https://web.archive.org/web/20070403125824, http://www.acl.be/News/sesquicentennial.htm.

VIS. 2008. Communications media. *Holy See Press Office*, January 24. http://visnews-en.blogspot.com/2008/01/communications-media-spreading-and.html.

Weber-Wulff, Debora. 2014. *False feathers*. Heidelberg: Springer.

Zenit. 2008. Benedict XVI: Media overstepping the mark. *Zenit*, January 24. https://zenit.org/articles/benedict-xvi-media-overstepping-the-mark.

Chapter 6
Exposition Plagiarism

Abstract Plagiarism in historiographical writing generates problems for readers, not the least of which is confusion concerning authorities. In a well-written, non-plagiarizing study, readers are presented with a clear separation of three kinds of authority: (1) the voices of authors of primary canonical texts whose works are cited; (2) the voices of exegetes in the secondary literature whose interpretations are analyzed, and (3) the voice of the author of the well-written study. In contrast, a plagiarizing historiographical study fails to demarcate clearly these three kinds of authority, and this failure vitiates the quality of that work. This chapter explores the phenomenon of *exposition plagiarism*, understood as plagiarism that involves the conflation of authoritative voices in historiographical writing.

Keywords Post-publication peer review · Historiography · Authority · Reception history

Historiographical works interpret canonical texts and trace their influence. A clear separation of three kinds of authority is central to successful historiographical writing: the authority of the authors of canonical primary texts, the authority of exegetes in the secondary literature, and the authority of a new author in producing a new work of exegesis. In a well-written historical study, readers will encounter primary texts (as cited), secondary literature (as referenced), and original interpretations (as argued) by the author of the historical study. Acts of academic plagiarism in published works of historiography conflate this three-fold order of authorities, creating problems for readers and weakening the conclusions of those works.

This chapter explores the various ways that plagiarism harms the quality of exegesis in published historiographical works. Such works fail to demarcate clearly these three kinds of authority, and this failure limits the ability of those works to advance a field with genuine contributions to knowledge. By misleading readers and creating problems in the downstream literature, such plagiarizing works corrupt a field rather than advance it. A recently published monograph that contains suspected plagiarism of various kinds provides the opportunity of a case study for exhibiting how plagiarism vitiates the quality of historiographical research. The following analysis is a detailed exercise in post-publication peer review that sets forth plagiarism varieties particular to historiographical writing by identifying and analyzing suspected examples of them.

The monograph was published in 2018 as volume 21 in the book series "History, Philosophy and Theory of the Life Sciences" issued by the academic international publisher Springer Nature. Overseen by team of 41 scholars from institutions around the world, the series presents "cutting-edge research" and "novel ways of tackling philosophical issues" in the area of the life sciences (Springer, n.d.). The author of record for the monograph—hereafter, "G."—proposes to explore "the rise of biology as a unified science in Germany" in authors from the mid-eighteenth to early-nineteenth centuries (G. 2018a: xv). Titled *Vital Forces, Teleology, and Organization: Philosophy of Nature and the Rise of Biology in Germany*, (hereafter, *VFTO*) the study covers German philosophers including Immanuel Kant and G. W. F. Hegel as well as many lesser-known theorists in the life sciences from that period (G. 2018a). The book appears with the author of record's institutional affiliation, and the author of record has published a dozen scholarly articles in English and Italian prior to the appearance of *VFTO*.

In the foreword written by a sponsoring scholar, *VFTO* is touted as "an important milestone for understanding how biology came about as an independent science" (Duchesneau 2018: v). Most published reviews of the book have been equally favorable, with both established figures in the field and new scholars providing their imprimatur through glowing reviews. In the flagship journal of the International Society for the History of Philosophy of Science, John Zammito characterizes G.'s book as "an excellent monograph" (Zammito 2018: 498). More importantly, he lauds that G. has presented "entirely appropriate stances on each and every one of the historiographical and philosophical issues at stake in this topic" (ibid.). Whether such a broad claim about historiography is sustainable in light of the evidence to be discussed here remains to be seen. In the analysis that follows in this chapter, G.'s allegedly appropriate stances on select historiographical and philosophical issues are called into question. Zammito then characterizes G. as "a very valuable new ally" in promoting views consistent with his own (ibid.). In a review in *Notre Dame Philosophical Reviews*, Georg Toepfer concludes that "[G.]'s book has great merits in providing a thorough analysis of some much-neglected thinkers of the Romantic school of *Naturphilosophie*" (Toepfer 2018). In another account published in *Philosophy in Review*, readers hear that the book "substantially contributes to our understanding of the emergence of biology during that period" (Kabeshkin 2019: 71). In a lengthy review in *Verifiche*, Daniele Bertoletti concludes that the book "is a useful and effective text [è un testo utile ed efficace]" (Bertoletti 2018: 317). None of the published reviews to date has raised any questions about the methods of G.'s study, and the general flavor of reviews has been commendatory. One brief outlier notice in *Berichte zur Wissenschaftsgeschichte* found the project of the book unconvincing but did not identify any particular methodological problems (Kanz 2018: 302–304). Even though the book has been published recently, it already has received generally positive citations in the downstream literature.

6 Exposition Plagiarism   105

Despite the litany of endorsements, it is arguable that the conclusions of *VFTO* are only as strong as the quality of evidence presented in favor of them. In any work of historiography, the quality of evidence requires an accurate interpretation of primary texts and a veridical presentation of the tradition of interpretation, especially as expressed in the relevant secondary literature. If the basic evidence is plagiarized or inadequately credited, then the quality of a historiographical study is compromised.

6.1 Entirely Unattributed Texts

The deficient attribution of sources in plagiarizing scholarly works admits of degrees of gravity. The most severe instances involve the absence of any attribution at all. In *VFTO*, there appears to be a range of deficient attribution to sources throughout the book. On one extreme are those instances where a researcher's words are used as an undisclosed source text, with no attribution given anywhere in the book. That is, there are no quotation marks, footnotes, or in-text citations, to signal to the reader that in some way that the words originate elsewhere. When the words of an undisclosed source text appear without any attribution at all, a finding of apparent plagiarism should be uncontroversial. In the first chapter of *VFTO*, on the second page of the book, an undisclosed source accounts for most of that page. Table 6.1 presents a selection of page 2 in the first column, with the suspected undisclosed source text in the second column, and the overlap between the two is highlighted.

The passage begins with the author of record—G.—using first-person narrative to introduce the account, saying "I begin my analysis [...]." Whether the analysis belongs uniquely to G. seems disputable. This opening locution suggests to readers that G. is the origin of the passage that follows, which provides an overview of the experiments and life work of seventeenth-century Genevan naturalist Abraham Trembley. The passage discusses Trembley's scientific discoveries on the regenerative properties of the hydra and considers how they were developed and disseminated. The apparent unattributed source text for this account, however, is a previously published article by Mary Sunderland that appeared first in 2007 and later in revised form in 2015 in *The Embryo Encyclopedia*, a peer-reviewed open access resource. There is no reference to Sunderland's work anywhere in the body of G.'s monograph, nor is her work listed in the bibliography at the end of the book. One must conclude that G. has taken the substance of Sunderland's article. A footnote reference in G.'s version of the text directs the reader to three works of scholarship by John Baker, Sylvia and Howard Lenhoff, and Marc Ratcliff—works identified in Sunderland's reference list at the end of her article—but none of these is the apparent source text. The presence of these three references in the note suggest that G. is providing a summary of the best of current literature for readers, but the hidden source is nearly all from Sunderland.

Much of the text appears to be verbatim and near-verbatim repetition from Sunderland's encyclopedia article. There is some paraphrase, and some of the text is moved around, but there is comparatively little that is not found in Sunderland. (Even the

Table 6.1 Sunderland 2015 and an undisclosed source text for G. 2018a

G. 2018: 2.	Sunderland 2015
I begin my analysis with a reference to the well-known experiments conducted by Abraham Trembley (1710–1784) on the green hydra in the first half of the eighteenth century.¹[Baker (1954), Lenhoff (1986), Ratcliff (2004).] In fact, the remarkable regenerative features of the small hydra that Trembley discovered caused many to question the standard beliefs about generation. Uncertainty about how to classify the hydra—was it an animal or plant?—convinced Trembley to conduct a series of experiments concerning its response to being damaged and its ability to regenerate. In his first observations, Trembley noticed that the polyp moved in a step-by-step way, much like an inchworm, which suggested it had an animal nature. He was therefore surprised to see that, when he cut its body in two halves, each half of the polyp regenerated, plant-like, into a complete new body. After detailed examination, Trembley finally concluded that there was no difference between the newly regenerated polyp and one that had never been cut. The publication of the results of this work had a revolutionary impact on the international scientific community. Trembley wrote his experiments first in a letter to René Antoine Ferchault de Réaumur (1683–1757), whom he had met through his cousin, Charles Bonnet (1720–1793). Réaumur was so excited by the results that he immediately announced them to the Paris Academy of Sciences. By the time Trembley's discoveries appeared in print, in volume 42 of the *Philosophical Transactions* of the Royal Society, most of the scientific community was already familiar with them. Some had already replicated his experiments and others had investigated the same phenomena in other organisms. Charles Bonnet, for instance, performed similar experiments on worms and published his discovery that some of them displayed the ability to regenerate alongside Trembley's paper. This support from the scientific community and wide confirmation of his results led to Trembley's election to the Royal Society in 1743. The following year, he published the *Mémoires, pour servir à l'histoire d'un genre de polyp d'eau douce, à bras en forme de cornes* (1744), which presented his experiments and observations in their entirety.	Abraham Trembley (1710-1784) Abraham Trembley's discovery of the remarkable regenerative capacity of the hydra caused many to question their beliefs about the generation of organisms. [...] Trembley was unaware of the polyp's identity and began a series of experiments to determine whether it was an animal or a plant. [...] During his observations, he noticed that the polyp moved in a step-by-step way much like an inchworm, which suggested that it was an animal. He was therefore surprised to see that each half of the polyp regenerated into a complete new polyp, which suggested that it was a plant. After detailed observation he concluded that there were no differences between the newly regenerated polyp and one that had never been cut. [...] Trembley wrote about his results in a letter to René Antoine Ferchault de Réaumur, whom he had met through his cousin Charles Bonnet in 1741. Réaumur was so excited by the results that he immediately announced them to the Paris Academy of Sciences. By the time Trembley's discoveries appeared in print in volume 42 of the Philosophical Transactions of the Royal Society most of the scientific community was familiar with his polyps. Some had already replicated his experiments and others had investigated the phenomena in other organisms. Charles Bonnet performed similar experiments in worms and published his discovery that some worms displayed the ability of regeneration alongside Trembley's paper. Trembley also published *Mémoires, pour server à l'histoire d'un genre de polyps d'eau douce, à bras en forme de cornes*, which presented his experiments and observations in their entirety. The support from the scientific community and wide confirmation of his results led to Trembley's election to the Royal Society in 1743.

dates given for figures René Antoine Ferchault de Réaumur and Charles Bonnet may originate in Sunderland's text, as in her encyclopedia entry proper names are hyperlinked to other entries on those figures and the dates appear in the titles of those linked articles.) To the reader it appears that G. is the authority setting forth the scientific contributions of Trembley, but the parallel in Table 6.1 shows that Sunderland's voice is expropriated without credit. This early part of *VFTO* appears to serve as a proxy through which readers unwittingly access Sunderland's scholarship.

Some might be tempted to downplay this apparent undisclosed dependency on Sunderland's work by rationalizing that the passage in question concerns generally uncontested factual or historical elements of Trembley's life rather than innovative exegetical analyses of primary texts or a novel estimation of the contributions of Trembley's scientific contributions. The official overview of the encyclopedia explains that "the descriptive articles in the *Encyclopedia* are written, peer reviewed, and edited to be factual, non-evaluative" (The Embryo Project Encyclopedia 2007). Nevertheless, the encyclopedia article on Trembly has an author of record—Sunderland—and her voice as author has been replaced with the republication of her work under different authorship.

Those who hold that a factual source text like an encyclopedia entry can be appropriated without committing plagiarism might be more interested in considering a second instance of an apparent dependency of a passage in G.'s monograph on an undisclosed source. A few pages later in this same first chapter of *VFTO*, one finds the beginning of a detailed exegesis of Caspar Friedrich Wolff's *Von der eigentümlichen und wesentlichen Kraft der vegetabilischen sowohl als auch der animalischen Substanz*, a 1789 work that attributes an "essential force" to plants and animals. Table 6.2 presents a selection of G.'s discussion alongside the apparent undisclosed source text.

The apparent source text for G.'s passage is a monograph published 37 years earlier by Shirley Roe. There are no quotation marks or in-text citations to indicate to the reader that the passage has already substantially appeared in print decades earlier under different authorship. Another portion of Roe's book was accurately cited four paragraphs earlier by G. in the book, but no indication is given here to disclose any dependency on Roe's discoveries. To the reader, the account presented by G. appears to be an original textual analysis of Wolff's work, beginning with the words "In this text [...]." This selection also contains what appear to be original English translations by G. of two passages from Wolff's 1789 work, as the footnote references are to the eighteenth-century German edition. But the English translations are the same verbatim selections as those translated by Roe in her work: even the idiosyncratic ellipses that Roe included in her quoted translation of Wolff are preserved in G.'s text.

Table 6.2 G. 2018a and Roe 1981 on C. F. Wolff's notion of essential force

G. 2018: 6–7	Roe 1981: 115–116
In this text Wolff upholds a simple model for how the essential force operates: in living bodies, he contends, similar substances attract one another, whereas different substances repel one another. On the basis of this phenomenon, Wolff claims, one can explain all vegetative activities: in nourishment, for example, liquids are brought to the different parts of the plant or animal, each part attracting material that is similar to it, which can be therefore used for growth and repair. Attraction is produced by the fact that two substances are similar in nature and is caused by the presence of the essential force in both the organism's nourishing liquid and its various parts. Through repulsion, a solidified part secretes material that is dissimilar to it, which solidifies to become a new structure and grows by attracting material similar to itself via the nourishing liquids. Wolff argues again, as he did in earlier works, that mechanical causes (for example the pumping of the heart after it is formed) influence vegetative activities but do not cause them. Their cause is the essential force, which "must be peculiar to this plant and animal substance, because no material other than plant and animal substance is nourished, vegetates, or reproduces its kind. Moreover, because the whole life of plants [and animals], their nutrition, growth, vegetation, and reproduction, rests upon it, one can call it a characteristic of the essential force. For where this force is absent, all vegetable process cease."[13][Wolff (1789), 66.] In other words, a body cannot be alive without the essential force. Nevertheless, Wolff does not understand this as a form of animism and therefore underscores the difference between himself and Stahl: "this characteristic essential force appears to be that, if I do not err [...], whose existence Stahl very certainly recognizes, but which he incorrectly attributed to the soul. It consists in nothing further than a particularly defined kind of attractive and repulsive force." The essence of life, Wolff contends, need not be attributed to a soul but to an attractive and repulsive force.	Wolff proposes a very simple model for how the essential force operates. In living organisms, he contends, like substances attract one another, whereas unlike substances repel one another. On the basis of this phenomenon, Wolff claims, one can explain all vegetative activities. In nourishment, for example, liquids are brought to the different parts of the plant or animal, each part attracting out material that is similar to it, which can therefore be used for growth and repair. The attraction is produced by the fact that the two substances are similar in nature and is caused by the presence of the essential force in both the nourishing liquid and the parts of the organism [...]. A solidified part secretes, through repulsion, material that is dissimilar to it. This then solidifies to become a new structure and can grow by attracting material similar to itself from the nourishing liquids. [...] He argues again, as he did in his early works, that mechanical causes (for example, the pumping of the heart after it is formed) influence vegetative activities but do not cause them. [...] The real plant is alive because of the essential force, which "must be peculiar to this plant and animal substance, because no material other than plant and animal substance is nourished, vegetates, or reproduces its kind. And because, moreover," Wolff concludes, "the whole life of plants [and animals], their nutrition, growth, vegetation, and reproduction, rests upon it, one can call it a characteristic and essential force. For where this force is absent, all vegetable processes cease" (p. 39). The structure of the organism cannot be alive without the essential force. Wolff again distinguishes his force from the vitalism of Stahl, even more explicitly than he had earlier in his dissertation. "This characteristic and essential force," he asserts, "appears to be that, if I do not err,...whose existence Stahl very certainly recognized, but which he, incorrectly I think, attributed to the soul. It consists in nothing further than a particularly defined kind of attractive and repulsive force" (1789:42): The essence of life, Wolff contends, need not be attributed to a soul but can, rather, depend upon an attractive and repulsive force.

The interpretation of Wolff's text, the selection of passages, and the translation are entirely unoriginal, having substantially appeared in print under Roe's name decades earlier. Roe's voice is unacknowledged, and readers are likely to be deceived into thinking that they are encountering original historiographical and translation work here. That is, readers mistakenly will assume the text expresses the authority of G. as author presenting a new account, rather than the established authority of Roe whose analyses are already published.

To be sure, there are some minor differences between Roe's 1981 text and G.'s 2018 text in the selection in Table 6.2. Some changes are merely synonym substitutions, but they are trivial. There are several other substitutions and additions, but overall the two passages are substantially identical. Roe's authority appears to be eclipsed with the lack of attribution in G.'s passage, and readers are apparently misled about the source of the analysis of Wolff's text, the selection of two extracts from the eighteenth-century work, and their translation into present-day English. In other words, the passage is almost entirely unoriginal as it appears in G.'s work.

6.2 Deficient Attribution

The previous section featured two examples from G.'s monograph where the apparent source texts were undisclosed to the reader. The traditional norms of using quotation marks, in-text citations, extracting, and the like were not employed in a way that would indicate to the reader that the passages have on the whole already appeared in print under different authorship. The authority of both Sunderland and Roe as established contributors to the secondary literature is apparently lost to the readership of G.'s monograph, who encounter the passages as if they were original. A deficient attribution of sources, however, occurs when an author of record gives some—though inadequate—reference to the source material. The inadequacy of the reference does not allow the reader to discern from the text itself that the words originate elsewhere, even though the source text appears in some understated or tangential way within the appropriated words. Table 6.3 presents an example of suspected deficient attribution, presenting a short passage from *VFTO* alongside the apparent source text. Again, this passage occurs in the first chapter of G.'s monograph in a discussion of Wolff.

This passage from *VFTO* appears to be largely composed of texts extracted from an abstract found at the beginning of a 2007 book chapter by Jean-Claude Dupont from the proceedings of a conference on Kant. The subject matter is Wolff's relationship with his contemporaries. There are some modifications, but the major claims—and the way they are expressed—are nearly identical to the apparent source text. For example, the expression "Wolff minutely analyzed" has been modified to "Wolff meticulously analyzed."

Two features stand out. First, G. provides a reference to Dupont 2007, the apparent source text, but only as the third item in the footnote that also references works by Georges Canguilhem and Shirley Roe. To the reader, it appears that the footnote is referencing the three authors in support of the specific claim that Wolff's 1768 work

6.2 Deficient Attribution

Table 6.3 Dupont 2018's abstract and G. 2018a

G. 2018: 3–4	Dupont 2007: 37, 43
In 1768, recognizing the uselessness of further theoretical argument with his opponents, Wolff found himself at an impasse in advancing his views about formative causes – though he would return to these later in discussion with Blumenbach – and instead produced a technical work intended to equal Haller's detailed treatise on the heart: *De formatione intestinorum* (1768). In this text, after summarizing his ideas about generation, Wolff meticulously analyzed the development of the digestive system. Although very technical, this treatise has been unanimously recognized as the first great text of modern embryology.[5][Canguilhem (1962), Roe (1981), Dupont (2007)] I will not summarize all of Wolff's observations concerning the envelopes of the embryo and the digestive system, which can be found elsewhere[6] [Dupont and Perrin (2003)] but will only stress [...].	In 1768, the German embryologist Caspar Friedrich Wolff (1733–94), recognizing the uselessness of further theoretical argument with Bonnet and Haller on invisibility, felt at an impasse in advancing his views about *vis essentialis* and formative causes, though he would return to these later in his discussions with Blumenbach. For the moment, it was necessary to produce a work whose professionalism would equal that of Haller on the heart: hence he produced *De formatione intestinorum*. After summarizing ideas about generation developed in his medical thesis of 1759 (*Theoria generationis*), Wolff minutely analyzed the development of the internal organs, mainly the digestive system. Recounting the genesis of this work, my chapter suggests that—[...]—it appears to be the first great text of modern embryology. // I shall not summarize all of Wolff's observations concerning the envelopes of the embryo and the digestive system here but shall emphasize [...].

De formatione intestinorum was "the first great text of modern embryology." The reader is likely not to conclude that the words before and after this claim largely overlap with text found in the third of the three referenced works in the note. If someone were to argue that the passage cannot be a case of suspected plagiarism since the source text is obliquely mentioned, the proper response would be to point out that source text is indeed mentioned but is not identified *as the source text*. The reference is not in the right place and it is not presented in a way that plausibly manifests to the reader that the bulk of the surrounding text before and after the note appears to originate in that text. Furthermore, one could ask why works by Roe and Canguilhem also are identified in the note, prior to the reference to Dupont. In short, the reader has no reason to suspect the words in G.'s passage here are not original, given the lack of quotation marks and the oblique mention of the source text. The example of suspected plagiarism in Table 6.3 appears to satisfy the conditions described in research integrity literature as *pawn-sacrifice plagiarism* (see Weber-Wulff 2014: 10–11).

The second significant feature of this passage and its apparent source is the use of the first person in both. In *VFTO* one finds, "I will not summarize all of Wolff's observations concerning the envelopes of the embryo and the digestive system [...]" and in Dupont one finds, "I shall not summarize all of Wolff's observations concerning the envelopes of the embryo and the digestive system [...]." The "I" in *VFTO* is really the "I" from Dupont, not G., but this fact is shielded from the reader. The use of the first person, the absence of quotation marks, and the burying of the reference to the source text behind two other references in the footnote jointly create the illusion that the passage in *VFTO* is original, when it fact it has already appeared in print under different authorship. The reader is at a disadvantage in assessing this text; she or he is likely to assume that G. is the genuine author rather than Dupont, whose contribution has been radically understated.

6.3 Incorrect Attribution

In addition to suspected instances of no attribution to a source text (Tables 6.1 and 6.2) and deficient attribution (Table 6.3), one can consider whether there are cases of suspected incorrect attribution. The second chapter of G.'s volume considers the contributions of "Göttingen School" to the development of the notion of vital force. The figures examined in detail include Albrecht von Haller, Johann Friedrich Blumenbach, and Carl Friedrich Kielmeyer. Table 6.4 presents a short selection from *VFTO*'s discussion of Haller, and at the end one finds a general footnote citation to a 1990 book by Maria Teresa Monti.

As the right column of Table 6.4 shows, however, the passage appears to derive immediately from an altogether different book: the passage closely re-presents text without reference from the first page of the Introduction to a 2005 book on Haller by author Hubert Steinke. The terminal footnote to Monti in G. 2018a conceals the fact that the text has already substantially appeared in print elsewhere by sending the reader to specific pages (45–77) of a different book rather than to the apparent source text by Monti.

Even if Steinke were to have been named in the footnote, instead of Monti, such an attribution would still have been insufficient to disclose to the reader the apparent dependency of the sentences on Steinke's book, since there are no quotation marks or extract quoting around the verbatim words. The minimum standards of attribution for scholarly writing are apparently not being met. The reader cannot discern—at least from the text found in *VFTO* itself—whose authority appears in many of its sentences and paragraphs. The absence of quotation marks around passages that can be found in previously published works by other authors leads readers to the default position that G. is the author of these sentences. The authority of G. as author of the book supplants the authority of various authors of the previously published secondary literature whose words appear without credit in *VFTO*.

Steinke's sentences have been modified slightly. As the passage appears in Steinke's introduction, the sentences describing Haller's experiments, conclusions,

Table 6.4 Incorrect attribution and no quotation marks

G. 2018: 36	Steinke 2005: 7
Haller's *De Partibus Corposis Humani Sensilibus et Irritabilibus* claims to have proven by means of animal experiments that only muscular fiber possesses the ability to contract. Haller calls this irritability and finds it responsible for movement. Haller clearly distinguishes this property from sensibility, which is responsible for sensual impressions and inherent only to nerves. He thereby challenges the traditional mechanical – mainly Boerhaavian – framework for animal physiology on three main points. First, by postulating a force inherent to the muscular fiber that is independent of the nerves and the soul. Secondly, by separating, both conceptually and physically, the properties of movement and sense perception. Third, by establishing a strict correlation between structure and function, not at the level of elementary particles but at the level of compounding structures, i.e. the muscular and nervous fibers.[9] [Monti 1990, 45–77.]	His orations *De Partibus Sensibilibus et Irritabilibus* [...] claimed to have proven by animal experiments that only the muscular fibre possesses the ability of contraction, which he called irritability and which was responsible for movement. From this property he strictly distinguished sensibility, responsible for sensual impression and inherent only in the nerves and the parts furnished with nerves. Thus he challenged the traditional, mechanical – mainly Boerhaavian – model on three main points. First, Haller postulated a force inherent in the muscular fiber and independent of the nerves and the soul. Second and partly as a result of this, he separated – conceptually and physically – the two properties of movement and sense perception. Third, and again in part resulting therefrom, he established a strict correlation between structure and function, not on the level of corpuscules or elementary particles, however, but on the level of compound structures, ie. the muscular and nervous fibres.

6.3 Incorrect Attribution

and motivations are written in the past tense. In *VFTO*, however, the verbs of these sentences have been changed to the historical present, so there is a shift in the way the verbs appear between the earlier and later texts (e.g., claimed→claims). A mere change of verb tense in an existing passage is insufficient in itself to produce an original historiographical interpretation of Haller's contributions to the development of the notion of vital force; the presentation is simply the same as Steinke's with a minor change to the form of the predicates. To the reader it appears that G. is engaging the work of Haller and providing an original analysis, but instead the evidence here suggests that G. is really appropriating the work of an unidentified secondary source (Steinke), merely changing the tense of the main verbs, and furthermore positing an oblique and irrelevant reference to Monti's book. There are other changes, but they largely involve synonym substitutions or equivalent phrasing. These changes do not affect the historiographical content, but they are merely minor alterations to the form. The overlap between the two passages can still be described as verbatim and near-verbatim identity.

In the next paragraph of *VFTO*, after the passage presented in Table 6.3, the discussion turns to Haller's experiments with Johann Georg Zimmermann, and a footnote directs the reader to pages 49–92 of Steinke's book. This reference does not afford credit for the previous paragraph, since the reader is directed to a discussion on a different topic 35+ pages away from the apparent source from the first page of the Introduction of Steinke's book. G. creates an illusion for the reader: the correct literature is being mentioned, but not in an accurate way, and not in a way to allow the reader to know of the verbatim and near-verbatim overlap. By forgoing quotation marks and giving footnote references to the wrong work, the reader is presented with a distorted view of the secondary literature. What seems to be G. speaking as author in these passages is apparently a façade in which the authority of various genuine authors in the secondary literature has been eclipsed. The reader is thereby disadvantaged in learning the true relationship of the history of interpretation of Haller's work in the secondary literature.

6.4 Presenting Debates in the Secondary Literature

The preceding discussion has focused on examples involving a complete lack of attribution to sources (Tables 6.1 and 6.2), a deficient attribution where the source is mentioned amidst two other works (Table 6.3), and finally an attribution to the wrong source (Table 6.4). In all these cases, an absence of quotation marks for apparently appropriated verbatim text corrupts the presentation. Readers are likely to draw incorrect assumptions about who is speaking in the text, i.e., whose authority is endorsing the claims being made. The passages in Tables 6.1–6.4 each concerned the analysis of the scientific contributions of some eighteenth-century figure (i.e., Trembley, Wolff, Haller). The discussion of these examples disclosed that the historiographical analyses that seemed to be original in *VFTO* had already substantially appeared in print, and attribution to the apparent sources was either lacking or inadequate. An

Table 6.5 Ginsborg's summary of her position as repeated in G. 2018a

G. 2018: 16	Ginsburg 2013
Ginsborg has attempted to resolve this problem by appealing to a conception of purposiveness as normativity, arguing that organized beings can be regarded as subject to normative standards without implying that they were in fact designed according to those standards [...].	Ginsborg (2001) attempts to resolve the problem of coherence by appeal to a conception of purposiveness as normativity (see Section 3.1 above), arguing that organisms can be regarded as subject to normative standards without supposing that they were designed to accord with those standards.

essential requirement of reliable historical studies is not only a veridical handling of the contributions of figures of long ago, but also an accurate presentation of the works of the secondary literature. There is evidence that even the way controversies in the secondary literature are set forth in *VFTO* creates further infelicities for the reader.

Table 6.5 displays a very short passage in which G. appears to summarize the contribution of Hanna Ginsborg to a debate on Kant's notion of teleology. The difficulty is that what appears to be G.'s original evaluation of Ginsborg's position is really Hanna Ginsborg's own summary of her 2001 book chapter that is described in an entry she wrote in *The Stanford Encyclopedia of Philosophy*. That is, G.'s passage here seems to be substantially repeating Ginsborg's autobiographical summary of her own contribution to a debate on Kant's notion of teleology.

As author of the entry in the encyclopedia, Ginsborg discusses her own contributions in the third person. A reader of the entry will see that Ginsborg as author of the entry is discussing her work as contributor to a particular debate about Kant. When her words appear in *VFTO*, however, G. appears to be the one providing a critical analysis of the secondary literature and judging the content of Ginsborg's contribution. Ginsborg's evaluation of her contribution is presented as G.'s evaluation. The authority shifts from Ginsborg to G. as readers think they are encountering G.'s critical appraisal of Ginsborg. Rather, they are receiving—unwittingly—Ginsborg's autobiographical account of her views on Kant as she expresses them in an entry in *The Stanford Encyclopedia of Philosophy*, a work not identified in *VFTO*.

This suspected use of a secondary author's own words in the guise of an original critical analysis of the literature is not a unique case. One of the primary projects of *VFTO* is to critique the work of Timothy Lenoir, and this task requires some discussion of how G.'s position differs from that of Lenoir. The summary of Lenoir's position found in *VFTO*, however, presents Lenoir's words often without quotation marks, giving the appearance that G. is offering an original, critical exposition and summary of Lenoir's position. There are references to Lenoir's work throughout, but since there are no quotation marks around many verbatim selections, readers are lead to believe that they are encountering the authority of G. as interpreter rather the authority of Lenoir as author. In short, Lenoir's autobiographical authorial voice is eclipsed. The lack of quotation marks creates the unusual situation where Lenoir's words are encountered by readers under the formality *as expressed by G.* rather than under the true formality *as expressed by Lenoir himself*.

To see how this phenomenon occurs, it is instructive to begin with an earlier version of a passage that G. published about Lenoir in a 2014 article in the journal *Studies in History and Philosophy of Biological and Biomedical Sciences* (*SHPBBS*).

6.4 Presenting Debates in the Secondary Literature   113

Table 6.6 Lenoir's summary of his position as presented in G. 2014

G. 2014: 18	Lenoir 1981: 190, 192, and 193
With the *Biologie* the "transcendental biology" of the Göttingen School had concluded its formative period. According to Lenoir, Treviranus succeeded in pulling together various aspects of the program that had been under intense discussion beginning in 1750 with Haller's translation of Buffon's *Histoire naturelle*. Treviranus synthesizes these conceptual elements into what he describes as a dynamic theory of organic nature, which he attempts to ground in an encyclopedic overview of biological research from mid-eighteenth century onward. Lenoir does not disregard the relationship between this conceptual framework and *Naturphilosophie*. Through careful reading of Schelling's early philosophy of nature in conjunction with the critical edition of his correspondence he deems it possible to show that while in Leipzig, in the period during which he devoted himself almost exclusively to acquiring background in natural science, Schelling concentrated on the works of the Göttingen School, particularly Lichtenberg, Blumenbach and Kielmeyer. Through direct personal contact with Pfaff and Eschenmaier, Schelling gained also an in depth knowledge of Kielmeyer's *Physik der Tierreichs*. In addition to these personal debts Lenoir identifies other intellectual ties between transcendental *Naturphilosophie* and speculative theories of nature. The concepts of *Einheit, Stufenfolge, Polarität, Metamorphose, Urtyp* and *Analogie* have been described as distinguishing elements of this approach to nature. The notion of ideal types was central to the Göttingen thought on comparative anatomy. Similarly, in the work of Kielmeyer the dynamic interaction of vital forces bears strong similarity to Schelling's notion of polarity, as well as the use made of that notion in the work of Nees von Essenbeck and Oken. Goethe's notion of metamorphosis is closely parallel to the modification of an original ground plan developed by the Göttingen biologists (Lenoir, 1981, 192–193). Surprisingly enough, after stressing the proximity between vital-materialism and *Naturphilosophie* Lenoir maintains that beside the strong similarities in key concepts of these approaches, there are still major differences in both their interpretation and significance. Most importantly, vital-materialism worked hard at remaining consistent with Kant's philosophy of organic nature,	With the *Biologie* of Treviranus the transcendental biology of the Gottingen School had concluded its formative period. In that work he had succeeded in pulling together all the various aspects of the program that had been under intense discussion since 1790, [...] beginning in 1750 with the translation of Buffon's *Histoire naturelle*. Treviranus synthesized these conceptual elements into what he described as the dynamic theory of organic nature, which he attempted to ground in an incredible encyclopedic overview of biological research since the mid-eighteenth century. // Through careful reading of Schelling's early works on *Naturphilosophie* in conjunction with the critical edition of his correspondence [...], it is possible to show that while in Leipzig, in the period during which he devoted himself almost exclusively to acquiring background in natural science, Schelling concentrated on the works of the Göttingen School, particularly Lichtenberg, Blumenbach, and Kielmeyer. Through direct personal contact with C. H. Pfaff and Eschenmaier, Schelling gained an in depth knowledge of Kielmeyer's "Physik des Tierreichs."[221] [...] In addition to these personal ties [...], there were other intellectual ties between transcendental *Naturphilosophie* and speculative theories of nature. The concepts of *Einheit, Stufenfolge, Polarität, Metamorphose, Urtyp*, and *Analogie* have been described as distinguishing characteristic elements of this approach to nature.[222] [...] The notion of ideal types, for instance, was central to Gottingen thought on comparative anatomy. Similarly, in the work of Kielmeyer the dynamic interaction of vital forces, [...] bears strong similarity to Schelling's notion of polarity, as well as the use made of that notion in the work of Nees von Essenbeck and Oken. The concept of metamorphosis central to speculative thought is closely parallel to the modification of an original ground plan that we have seen developed in the work of the Gottingen biologists. [...] There are indeed strong similarities in key concepts of these approaches, but there are major differences in both the interpretation and significance of these concepts [...]. It is most important to realize that the transcendental approach worked hard at remaining consistent with Kant's philosophy of organic nature.

Table 6.6 presents the passage from G.'s 2014 article with the apparent source text by Lenoir.

Toward the end of G.'s passage an in-text citation directs the reader to pages 192–193 of Lenoir's lengthy 1981 article. Yet the citation does not suffice as an attribution disclosing the verbatim and near-verbatim repetition of Lenior's own words. The lack of quotation marks, and the fact that the passage is also presenting verbatim and near-verbatim text from earlier in the work—page 190—means that readers are not in a position to assess G.'s passage accurately. What appears to be a critical analysis of Lenoir's position by G. is really an autobiographical analysis by Lenoir of his own position. There is some condensation of Lenoir's autobiographical account, as indicated by the ellipses in Table 6.6, but the common sentences still fall in the same order.

In *VFTO*, G. incorporated much of the content of his earlier 2014 article, but in edited form. The incorporation included an edited version of the passage shown in Table 6.6. The editing for the 2018 version lessens the overall appearance of dependency on Lenoir's book, but much of the overlap is identifiable. Table 6.7

Table 6.7 Lenoir's summary of his position as presented in G. 2018a

G. 2018: 127–128	Lenoir 1981: 190, 192, and 193
The "transcendental biology" of the Göttingen School, Lenoir claims, thereafter concluded its formative period, as Treviranus succeeded in consolidating various aspects of the program that had been under intense debate since the 1750s with Haller's translation of Buffon's *Histoire naturelle* and synthesizing them into a dynamic theory of organic nature. Lenoir does not disregard the relationship between Göttingen's conceptual framework and Romantic *Naturphilosophie*. Indeed, through careful reading of Schelling's early philosophy of nature, in conjunction with the critical edition of his correspondence, Lenoir in fact argues that while in Leipzig, a period during which Schelling devoted himself almost exclusively to acquiring background in natural science, he studied the works of the Göttingen School, especially Georg Christoph Lichtenberg (1742–1799), Blumenbach, and Kielmeyer. Through direct personal contact with Christian Heinrich Pfaff (1773–1852) and Karl August Eschenmayer (1768–1852), Schelling also gained in-depth knowledge of Kielmeyer's *Physik der Tierreichs*. In addition to these personal connections and influences, Lenoir highlights other intellectual ties between the two traditions, such as the concepts of *Einheit, Stufenfolge, Polarität, Metamorphose, Urtyp* and *Analogie*. Similarly, the dynamic interaction of vital forces in the work of Kielmeyer bears strong similarity to Schelling's notion of polarity, as well as to the use Nees von Esenbeck (1776–1858) and Oken make of that notion. Surprisingly, however, after stressing the proximity between Göttingen and Jena (the capitol city of Romantic *Naturphilosophie*), Lenoir maintains that, besides the strong similarities across key concepts of these approaches, there are major differences in the way they interpret and grant significance to those concepts. Most importantly, he argues that the Göttingen tradition worked hard at remaining consistent with Kant's philosophy of organic nature.	With the *Biologie* of Treviranus the transcendental biology of the Göttingen School had concluded its formative period. In that work he had succeeded in pulling together all the various aspects of the program that had been under intense discussion since 1790, [...] beginning in 1750 with the translation of Buffon's *Histoire naturelle*. Treviranus synthesized these conceptual elements into what he described as the dynamic theory of organic nature, which he attempted to ground in an incredible encyclopedic overview of biological research since the mid-eighteenth century. // Through careful reading of Schelling's early works on *Naturphilosophie* in conjunction with the critical edition of his correspondence [...], it is possible to show that while in Leipzig, in the period during which he devoted himself almost exclusively to acquiring background in natural science, Schelling concentrated on the works of the Göttingen School, particularly Lichtenberg, Blumenbach, and Kielmeyer. Through direct personal contact with C. H. Pfaff and Eschenmaier, Schelling gained an in depth knowledge of Kielmeyer's "Physik des Tierreichs."[221] [...] In addition to these personal ties [...], there were other intellectual ties between transcendental *Naturphilosophie* and speculative theories of nature. The concepts of *Einheit, Stufenfolge, Polarität, Metamorphose, Urtyp,* and *Analogie* have been described as distinguishing characteristic elements of this approach to nature.[222] [...] The notion of ideal types, for instance, was central to Gottingen thought on comparative anatomy. Similarly, in the work of Kielmeyer the dynamic interaction of vital forces, [...] bears strong similarity to Schelling's notion of polarity, as well as the use made of that notion in the work of Nees von Essenbeck and Oken. The concept of metamorphosis central to speculative thought is closely parallel to the modification of an original ground plan that we have seen developed in the work of the Gottingen biologists. [...] There are indeed strong similarities in key concepts of these approaches, but there are major differences in both the interpretation and significance of these concepts [...]. It is most important to realize that the transcendental approach worked hard at remaining consistent with Kant's philosophy of organic nature.

presents the selection of text as it appears in *VFTO*, alongside the relevant passage in Lenoir's 1981 chapter.

Synonym substitutions replace terms and expressions for various parts of speech, yet despite these changes, the dependency of G.'s account of Lenoir's work still appears extracted from Lenoir's account of his work. Readers seem to be unknowingly encountering the authority of Lenoir in presenting his own position when they read this passage in *VFTO*.

One reason why researchers who extolled *VFTO* in published reviews and endorsements without apparently discerning the particular instances of suspected plagiarism may be the two-stage procedure seen in the appropriation of Lenoir's article. If one compares Table 6.6 with Table 6.7, one will see that the degree of verbatim overlap with Lenoir's article is more evident in the earlier 2014 article version shown in Table 6.6 than in the later 2018 *VFTO* version shown in Table 6.7. When a plagiarizing text is altered with a second round of editing, there is greater distance between the source text and the twice-removed plagiarizing version. This phenomenon can be seen in another example involving the same 2014 article and another portion of *VFTO*. Table 6.8 presents a selection from the introduction by Keith Peterson to his 2004 English translation of Schelling's *First Outline of a System of the Philosophy*

6.4 Presenting Debates in the Secondary Literature

Table 6.8 Selections of Peterson 2004 in G. 2014 and G. 2018a

G. 2014: 17	G. 2018: 81	Peterson 2004: xxix–xxx
They differ from each other not only in their material composition, but in the relation and proposition of their constitutive forces among themselves. According to Schelling, a species is the expression of a certain proportion of primary functions. The continuity of organic functions throughout the animal kingdom forms a universal organism. Schelling establishes a comparative physiology of organic functions, which endeavors to establish the various degrees and proportions of essential forces belonging to different organisms. In fact, every organism displays a specific proportion of reproductive force, irritability and sensibility. It is primarily defined not by its external form, but by the particular proportion of forces active within it. Its form and organs will follow from the disposition of these forces, each one prominent in a different class. Plants, for example, have a preponderance of reproductive force while displaying no sensibility. Mammals have a preponderance of sensibility but can generate few offspring and they do not display regenerative features which are frequent in amphibians.	They differ from each other not primarily with regard to their material composition but in terms of the relative proportion of vital forces they display. Like Kielmeyer, Schelling thus develops a comparative physiology of vital forces whose aim is to establish the various degrees and proportions of each vital force in the animal kingdom. He defines every organism as a specific proportion of reproductive force, irritability, and sensibility. Every organism is defined not primarily by its external form but by the proportion of these forces active within it. Its form and organs follow from the nature and proportion of these forces. For instance, every organic being is suffused with all three, but plants have a prevalence of reproductive force while their sensibility is close to zero. On the other hand, in mammals sensibility is dominant but they produce few offspring; their reproductive force so narrows that they retain only the capacity to reproduce the organism itself through growth, assimilation, and maintenance.	One product is specifically different from another not only in its material composition, but in the relation and proportions of its constitutive forces among themselves [...]. On a larger scale, what a species expresses of the whole is [...] a certain proportion and intensity of primary organic functions. It is through the continuity of organic functions that the whole diversity of the natural world is connected and forms a single whole organism. [...] Schelling establishes a "comparative physiology" of organic functions. [...] [H]e endeavors to establish a science of the various degrees and proportions of essential powers that belong to all organic beings. [...] Schelling defines every organism as a permanent process by attending to its specific proportion of reproductive force, irritability, and sensibility. Every organism is defined, not primarily by its external form [...] but by the particular proportion of forces active within it. All organic beings are suffused with all three powers, and yet plants, for example, have a preponderance of reproductive force while sensibility in them approaches zero. Mammals, in contrast, have a preponderance of sensibility but produce few offspring, their reproductive power narrows to a capacity to reproduce only the organism itself through growth, assimilation, and maintenance.

of Nature. Peterson's introductory essay is apparently an undisclosed source text for a passage in G.'s 2014 article in *SHPBBS* and then again in 2018 in *VFTO*. The first column presents a passage from G. 2014 with unattributed overlap with Peterson 2004 highlighted; the second column presents the same passage as it appears in edited form in the *VFTO* (again without attribution); and overlap is indicated with underlining. The third column presents the undisclosed passage from Peterson 2004.

In the 2018 version in *VFTO*, the apparent overlap of G.'s passage with Peterson's introduction still is evident, despite additional editing after the 2014 version. Some portions that don't overlap are still very close paraphrases or synonym substitutions for expressions found in the 2014 source. The consistent order of the verbatim and near verbatim parts also supports a finding that the passages in the 2014 article and 2018 book derive from Peterson's uncredited book. In neither the 2014 article nor *VFTO* is there any reference to Peterson's introduction or translation of Schelling's work. Furthermore, the several quotations in English of Schelling's work that are given in *VFTO* only reference a German edition, but the translation appears to be from Peterson's translation, again without any credit (G. 2014: 17, 2018a: 81–82; Peterson 2004: 51–53, 142).

6.5 Primary Texts in Philosophy

The preceding account has focused on examples exhibiting suspected cases of deficient attribution in the use of works from the secondary literature on eighteenth- and nineteenth-century biological theories. When encountering these examples in *VFTO*, the typical reader will believe that she or he is reading original work because the identity of the genuine authors has been suppressed in the absence of traditional forms of attribution. This particular problem is not the only failure of attribution in *VFTO*, however. As noted above, historiographical works should be judged in two important ways: (1) how well they handle the history of interpretation as reflected in the secondary literature; and (2) how well they handle primary canonical texts in the relevant field.

The philosopher Immanuel Kant receives much treatment in *VFTO*. In the first chapter, G. provides an account of Kant's views on the requirements of science as expressed in various works, including Kant's 1786 book, *Metaphysische Anfangsgründe der Naturwissenschaft*. Yet, what purports to be G.'s presentation of that work on the subject in question is remarkably close to Kant's own words, as they are rendered into English in a translation of that work by Michael Friedman in the series Cambridge Texts in the History of Philosophy. Table 6.9 exhibits the overlap between G.'s analysis and the text from Kant's own summary of his position in the Preface of *Metaphysical Foundations of Natural Science*.

The difficulty is that readers are encountering the words of Kant's preface to *Metaphysical Foundations of Natural Science* under the guise of a new interpretation of

Table 6.9 Kant's views on Kant in G. 2018a

G. 2018: 19	Kant 2004: 4–7 (trans. Friedman)
A rational doctrine of nature thus deserves the title of a natural science only if the fundamental laws therein are known a priori and not the mere result of experience. If the grounds or principles are merely empirical, as in chemistry, they carry no consciousness of their necessity. In this case, the knowledge involved does not merit the title of natural science. Accordingly, chemistry should be considered a systematic art rather than a science. Natural science instead derives its legitimacy from its 'pure' basis in the a priori principles of natural explanations. Indeed, explanation based on chemical principles always leave behind a certain dissatisfaction for Kant, because one can adduce no a priori grounds for these principles, which, as contingent laws, have been learned merely from experience. In Kant's view, although a pure philosophy of nature (i.e. that which investigates only the concept of nature in general) may be possible without mathematics, a pure doctrine of nature is only possible by means of mathematics: "in any special doctrine of nature there can be – in fact – only as much proper *science* as there is mathematics therein."[63][*Ivi*, 470.] Therefore, as long as there is no a priori law to explain chemical effects, chemistry can be nothing more than a systematic art or experimental doctrine – not a proper science.	A rational doctrine of nature thus deserves the name of a natural science, only in case the fundamental natural laws therein are cognized a priori, and are not mere laws of experience. [...] If, however, the grounds or principles themselves are still in the end merely empirical, as in chemistry, for example, and the laws from which the given facts are explained through reason are mere laws of experience, then they carry with them no consciousness of their necessity (they are not apodictally certain), and thus the whole of cognition does not deserve the name of a science in the strict sense; chemistry should therefore be called a systematic art rather than a science. [...] natural science must derive the legitimacy of this title only from its pure part – namely, that which contains the a priori principles of all other natural explanations [...] Hence, the most complete explanation of given appearances from chemical principles still always leaves behind a certain dissatisfaction, because one can adduce no a priori grounds for such principles, which, as contingent laws, have been learned merely from experience. [...] Hence, although a pure philosophy of nature in general, that is, that which investigates only what constitutes the concept of a nature in general, may indeed be possible even without mathematics, a pure doctrine of nature concerning determinate natural things (doctrine of body or doctrine of soul) is only possible by means of mathematics. And, since in any doctrine of nature there is only as much proper science as there is a priori knowledge therein, a doctrine of nature will contain only as much proper science as there is mathematics capable of application there. [...] So long, therefore, as there is [...] chemistry can be nothing more than a systematic art or experimental doctrine, but never a proper science.

6.5 Primary Texts in Philosophy   117

Table 6.10 Hegel's views on Hegel in G. 2018a

G. 2018: 119	Hegel 1977: 155–156 (trans. Miller)
In fact, often as we may find a thick, airy pelt associated with northern latitudes or the structure of a fish associated with water, "the notion of north does not imply the notion of a thick airy pelt, the notion of sea does not imply the notion of the structure of fish, or the notion of air does not imply the structure of birds."[7] [Hegel (1980), 146.]	[B]ut as often as we may find a thick, hairy pelt associated with northern latitudes, or the structure of a fish associated with water, or that of birds with air, the Notion of north does not imply the Notion of a thick hairy pelt, the Notion of sea does not imply the Notion of the structure of fish, or the Notion of air that of the structure of birds.

Kant. The authority of Kant as author of the primary text—a canonical text in the discipline—is erased and readers are lead to believe that they are encountering the authority of G. as expositor of Kant. G.'s analysis is sprinkled with occasional quotations from Kant that are marked off in the traditional way with quotation marks and a footnote citation, and this practice further supports the illusion that the parts not quoted are original expositions by G. of Kant's work. The exchange of authority—G. for Kant—is invisible to the reader. The modifications to Kant's words are often minor, yet nevertheless what the reader encounters for the most part here is Kant's own writing under the façade of an original, new exposition. The short example presented in Table 6.9 consists of texts taken from sentences extracted from pages 4–7 of Friedman's translation of Kant's preface to the work. In addition to being a case of apparent exposition plagiarism, the passage is a case of apparent compression plagiarism, since a lengthy passage is being distilled into a shorter one and presented under new authorship (see Chap. 3). As a historiographical contribution to the understanding of Kant, the exposition fails to be new, and the genuine authorship of it is not disclosed to the reader.

A similar yet less severe instance of this problem can be found in the treatment of Hegel in *VFTO*. In arguing that Hegel opposes Gottfried Reinhold Treviranus's position on the relationship of environment to the internal structure of animals, G. includes in his analysis a brief quotation from Hegel's *Phenomenology of Spirit*. The short passage is set forth in Table 6.10.

Several factors weaken the quality of G.'s short presentation here. First, as can be seen from the side-by-side comparison with Hegel's text, the quotation marks in G.'s passage show up rather late; the verbatim text from Hegel begins much earlier. That is, what appears to the reader to be G.'s introduction of Hegel's text is verbatim text from Hegel, with the exception that Hegel's adjective "hairy" has been errantly presented in *VFTO* as "airy," which changes the meaning of the text to some degree. More importantly, however, the use of the first-person plural pronoun "we" that appears in G.'s text is really the "we" as present in the English rendering of Hegel in the translation by A. V. Miller. Oddly, the footnote reference to Hegel directs the reader to a 1980 edition of the German text of Hegel's work, but the translation here, and the words that precede the translation, appear to be directly from Miller's translation. No acknowledgment is given for the use of Miller's translation, and it is not listed in the bibliography.

In an additional oddity, on the next page of *VFTO* G. points out Hegel's unusual tendency to use texts of others without giving citations, noting "Hegel quotes a passage from Kielmeyer's 1793 lecture without explicit reference, as often occurs

Table 6.11 Smith 2011 on Leibniz's views on animal bodies in G. 2018a

G. 2018;: 17–18	Smith 2011: 14–15
Leibniz understands the animal body as a divine machine, distinguished from the ordinary products of human artifice through its infinite complexity and consequent indestructibility. […] It is thus not surprising that Leibniz endorsed preformation: for him organs were designed by an omniscient creator and brought into existence, all together, at creation	Leibniz […] believes that the animal body is a natural machine or, which is the same, a divine machine whose infinite complexity and consequent indestructibility are enough to place it in a different ontological category from the ordinary products of human artifice. […] Instead, for him organs are designed by an omniscient creator for the execution of functions that are all brought into existence together at the creation

in his writings" (G. 2018a: 120). That G. would identify this practice is remarkable, since the apparent prevalence of the practice in *VFTO* substantially weakens the quality of historiographical scholarship in the book.

A lack of originality even mars the presentation of minor figures who are tangential to the account presented in *VFTO*. Table 6.11 gives one example from the very brief mention of Gottfried Wilhelm Leibniz, and the short discussion seems to extract key phrases from a 2011 book on Leibniz by Justin E. H. Smith.

G. appears to be repeating idiosyncratic expressions from the previously published book on Leibniz (e.g., "infinite complexity and consequent indestructibility"). There is no citation to Smith's book within the body of *VFTO*, but the book is listed in the bibliography.

6.6 Primary Texts in Biology

The apparent pattern of re-presenting the words of canonical writers as new analyses of their works is not limited to G.'s account of major eighteenth- and nineteenth-century philosophers. The practice also surfaces in *VFTO*'s coverage of the works of the period by major figures who are primarily known for their contributions in biology. The third chapter of the monograph, for example, considers various works by the German naturalist Lorenz Oken and highlights the value of each one in relation to G.'s project of tracing the establishment of biology as a modern science. The analyses of Oken's early works, however, does not appear to be entirely original. When discussing Oken's works from the early period from 1802–1805, G. appears to be extracting Oken's own account of those early works that Oken himself published several years later in the preface to his lengthy *Lehrbuch der Naturphilosophie* (1809–1811). But there is a twist: G. appears to be extracting Oken's own account of his early works not directly from the German text of the *Lehrbuch* but via Oken's account as rendered in an 1847 English translation of it by Alfred Tulk that was published under the title *Elements of Philosophy*. Table 6.12 presents a passage from G.'s account of Oken's early writings found in *VFTO* and the corresponding undisclosed apparent source from the first page of the Preface of Oken's work in the nineteenth-century English translation.

In the original apparent source text—*Lehrbuch der Naturphilosophie* (1809–1811)—Oken evaluates the significance of his early works, highlighting what he

6.6 Primary Texts in Biology

Table 6.12 Oken's view on Oken in G. 2018a

G. 2018: 85	Oken 1847: xi (trans. Tulk)
As a student, Oken had already composed the *Übersicht des Grundrisses des Systems der Naturphilosophie* (1802), a short writing strongly inspired by Schelling, in which he laid out his general research program for natural history. In this early work, Oken maintains that the animal classes are nothing other than an index of the sense organs and that they should therefore be arranged accordingly. Thus, strictly speaking, there are only five animal classes: *Dermatozoa*, or Invertebrates, *Glossozoa*, the Fishes, the first animals to appear with a true tongue; *Rhinozoa*, or Reptiles, which first exhibit a nose that opens into the mouth and inhales air; *Otozoa*, or Birds, in which the ear for the first time opens externally; and *Ophthalmozoa*, or Thricozoa, in which all the sense organs are present and complete and the eyes are moveable and covered with two palepbrae, or lids. [...] In this essay, *Die Zeugung* (1805), Oken argues that all organic beings originate from and consist of vesicles or cells and that these vesicles are the infusoria (aquatic organisms at the boundary between the plant and animal kingdoms) from which all larger organisms fashion themselves.	The first principles of the present work I laid down in my small pamphlet entitled *Grundriss der Naturphilosophie, der Theorie der Sinne und der darauf gegrundeten Classification der Thiere*; Frankfurt bey Eichenberg, 1802, 8vo (out of print). I still abide by the position there taken, namely, that the Animal Classes are virtually nothing else than a representation of the sense-organs, and that they must be arranged in accordance with them. Thus, strictly speaking, there are only 5 Animal Classes: Dermatozoa, or the Invertebrata; Glossozoa, or the Fishes, as being those animals in whom a true tongue makes for the first time its appearance; Rhinozoa, or the Reptiles, wherein the nose opens for the first time into the mouth and inhales air; Otozoa, or the Birds, in which the ear for the first time opens externally; Ophthalmozoa, or the Thricozoa, in whom all the organs of sense are present and complete, the eyes being moveable and covered with two palpebrse or lids. [...] I first advanced the doctrine, that all organic beings originate from and consist of *vesicles* or *cells*, in my book upon Generation. (*Die Zeugung*. Frankfurt bey Wesche, 1805, 8vo.) These vesicles, when singly detached and regarded in their original process of production, are the infusorial mass, or the protoplasma (*Ur-Schleim*) from whence all larger organisms fashion themselves or are evolved.

believes to be the distinctive contribution in each. As the selection in Table 6.12 shows, Oken expresses himself in the first person, saying "I first advanced the doctrine [...]"; "I still abide [...]"; "I laid down [...]." When the autobiographical words of Oken are apparently appropriated by G., the first person is transformed to the third person (e.g., "Oken maintains [...]; "Oken argues [...]). There is a significant shift in authority as the words spoken by Oken are now presented as coming from G., and this shift is not apparent to the reader. Ideally, a reader should know that Oken—a canonical author in the history of biology at the turn of the nineteenth century—is the true speaker of these words, and his analysis of his own earlier writings carries a weight for students and scholars that is not identical to the authority of G. as the author of a new historiographical study. G. appears to transpose Oken's authority with his own in the presentation of the significance of Oken's early writings. The loss of original authority arguably lessens the quality of the exposition. There are some minor changes introduced by G. to the text, but these do not change the substance of Oken's autobiographical account of his earlier works. Older-sounding locutions have also been updated, as "wherein" and "whence" are changed to "which."

G. does give two quotations with proper citation from Tulk's translation when discussing the significance of *Lehrbuch der Naturphilosophie* elsewhere in the book. The issue here, however, is that the English version of the words of Oken in the *Lehrbuch* is apparently being used as an unacknowledged source for the analysis of *Übersicht des Grundrisses des Systems der Naturphilosophie*, a point not disclosed to the reader. Some historical inaccuracies follow from taking an autobiographical primary text from the early 1800s and presenting it as new critical analysis in the twenty-first century, since any discoveries made in the two century period will be missed. Present-day research, for example, has established that Oken's self-published work *Übersicht des Grundrisses des Systems der Naturphilosophie* did not appear in print in 1802, as Oken asserts in his autobiographical Preface in *Lehrbuch der*

Naturphilosophie, but it appeared in 1804, even though it was written in 1802 (see Bach and Neuper 2001: 251).

The apparent use of Oken's text here cannot be downplayed as a merely factual biblio-biographical presentation of Oken's early writings. Oken's critical—though brief—analyses of the significance of his prior writings is a valuable key for interpreters in coming to understand Oken's views on the development of biology at the time. The reader is disadvantaged by being denied the true source of the estimation of the merits of Oken's early writings. Furthermore, no credit is given to the nineteenth-century translator of Oken's work—Alfred Tulk—whose translation is apparently the direct source for G.'s passage presented in Table 6.12. The act of translation is not a negligible modification of a canonical text; every transformation of a text through translation should be noted in historiographical studies that seek to present the most reliable interpretations of texts and to examine how they have been received.

An accurate historiographical account would acknowledge these elements. To the reader, it appears that G. is presenting an original interpretation of the significance of Oken's early works, as depicted in Fig. 6.1.

In contrast, Fig. 6.2 illustrates the suspected transmission of text argued for here: Oken's 1802/4 work (1) is interpreted by Oken himself in his later 1809–1811 work (2), translated into English by Tulk (3), and then presented as an original interpretation by G. in 2018 (4). In other words: Oken's own interpretations of his early works,

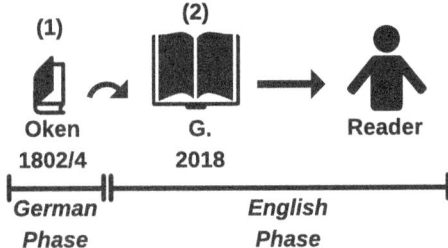

Fig. 6.1 How the interpretation of G. 2018a appears to a reader

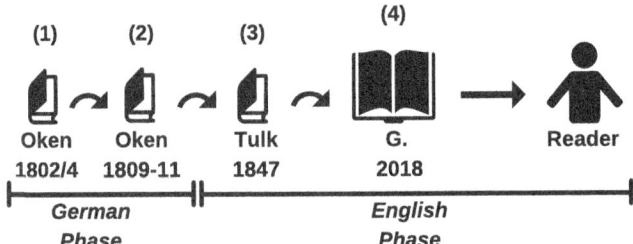

Fig. 6.2 The hidden transmission of Oken's text via Turk's translation

6.6 Primary Texts in Biology

put forth within a decade of their composition, have been translated into nineteenth-century English and then extracted with some modifications and presented in *VFTO* as if they were original twenty-first century interpretations.

As the reconstruction suggests, the reader is left unaware of part of the German phase of the transmission of the text (2) as well as part of the English phase (3), in addition to being unaware that the analyses of Oken 1802/4 (1) originate in an English translation (3) of Oken 1809–11 (2) rather than G. 2018a (4).

Another apparent instance where a canonical biological text is fashioned into a present-day critical interpretation of it involves *VFTO*'s treatment of Albrecht Haller's most famous work. In the second chapter of the monograph, G. discusses Haller's 1753 *De partibus corporis humani sensibilibus et irritabilibus*, which was issued in an anonymous English translation in 1755 as *A Dissertation on the Sensible and Irritable Parts of Animals*. In this work, Haller describes a variety of experiments he conducted on 190 animals for the sake of determining which parts of animals were "irritable," that is, disposed to contraction upon being touched, and those parts that are "sensible," that is, susceptible of conveying pain to the animal. G. presents an analysis of Haller's work, but much of the analysis appears to be selected from sentences taken from Haller's work in English translation. G. does include in his account some quotations from Haller's work with footnote citations, and he provides references throughout his account, but the lack of quotation marks in other parts occludes the extent of overlap with the source text. Table 6.13 offers a selection of G.'s extensive discussion of Haller's work.

The analysis offered by G. appears to consist largely of extracts from Haller's work. Much like Oken, Haller describes his experiments in the first person, again leading G. to transform sentences expressed in first person into the third person. So, for example, Haller's "After I was fully satisfied […]" becomes "After Haller was fully satisfied […]." What appears as twenty-first-century analysis of Haller's work seems to be an undisclosed anthology, compendium, or florilegium of texts from Haller's work. There can be value to disclosed works of that kind, but the lack of quotation marks around most of the verbatim text leaves the reader unable to discern that the words in many cases were expressed first by Haller in 1775 and translated in 1936, before appearing in *VFTO* in 2018. The presence of some cited quotations, as noted above, further generates the appearance that the author of record is professionally committed to the expected demarcation between the citation of a canonical text and a new analysis of it, but nearly all of the passage is directly from Haller's *magnum opus*. Again, the relevant authority is not clear to the reader, who is unable to discern that Haller's voice is being presented with inadequate acknowledgment.

There are some changes to Haller's words. Even in the cited quotation from Haller, the British English orthography of the anonymous translation has been modified: "endeavouring" and "fibre" are rendered as "endeavoring," and "fiber" in *VFTO*. In the unquoted material, there have been some minor synonym substitutions that do not change to substance of the passage.

Table 6.13 Haller's views on Haller in G. 2018a

G. 2018: 37, 38–39.	Haller 1936: 651–699
[A]fter Haller was fully satisfied with the certainty of this phenomenon that he began to address its cause: that there are nerves in the muscles but not in the tendons. […] As a result, Haller claimed to have identified the irritable parts of the human body and to what degree they were irritable. He concluded that the skin, cellular membrane, and fat are not irritable parts nor are the lungs, kidney, liver, or spleen, since they are composed of cellular substance. Irritability seemed to Haller "to be that which distinguishes the cellular fibers from the muscular," since the cellular membrane's capacity for irritability is precisely the same as that of the fibers of dead flesh: yielding to the touch, dimpling if pressed, and recovering once the pressure is removed. On the other hand, when irritated with a knife or corrosives in a living body, muscular fibers become shorter, and their extremities approach each other; then they are relaxed again, and these contractions and relaxations alternatively repeat for some time. On the other hand, the tendons are as devoid of irritability as they are of sensation; no irritation by knife or gentle corrosive can excite convulsion in them or produce any motion in the muscle to which the irritated tendon belongs. The ligaments, periosteum, meninges of the brain, and all the membranes composed of cellular membrane also lack irritability. Lacteal vessels contract and empty themselves upon being touched with oil of vitriol; what provs that they are considerably irritable is that, even if they are full at the time of the animal's death, they empty themselves and contract.[15][680–681] Pricking the bladder of a dog that was almost dead with a knife or needle frequently caused considerable spontaneous contraction. Muscles are irritable, since they all have at least one natural palpitation after death; they tremble and are alternatively contracted and relaxed. The esophagus contracts itself if irritated above the diaphragm. The stomach is highly irritable: when touched with a corrosive, the resulting impression immediately produces a long superficial furrow. The intestines, both large and small, are extremely irritable. By degrees, Haller proceeds to the most irritable organ of all, the heart, the cause of all motion in the human body. The heart is the most constructed for contraction and therefore, Haller believed, ought to be endowed with the greatest irritability. His experiments confirmed this to be so, especially in cold-blooded animals, where the heart is constantly irritable – much more than even the intestines.[16][687] Collecting all of Haller's experiments together, it appears that "there is nothing irritable in the animal body but the muscular fiber and that the faculty of endeavoring to shorten itself when we touch it is proper to the fiber. From the same experiments it likewise follows, that the vital parts are the most irritable." The diaphragm frequently moves after all the other muscles have ceased, the intestines and stomach move still longer, and the heart continues its motion after all the other parts are quiet.	[662–3] After I was fully satisfied of the certainty of the above event, I easily discovered the cause, *vis.* that there are nerves distributed to the muscles, but not to the tendons. [679] I have discovered which are the irritable parts of the human body, and in what degree they are so. I have excluded the skin. The cellular membrane and the fat. which the oil of vitriol devours so greedily, is generally agreed to be void of motion, at least a gentle irritation does not affect it. So neither the lungs (tho' violent acids constrict them) the liver, the kidneys, nor the spleen, have any irritability; because they are composed of the cellular substance, which, as I said just now, is not endowed with it, and of vessels which are equally void of irritability. This character of irritability seems to me to be that which distinguishes the cellular fibres from the muscular. […]. The disposition of the cellular membrane to irritability, is precisely the same with that of the fibres of dead flesh; it yields to the touch, dimples if you press it, and recovers itself if the pressure is removed. If you cut it, the fibres retract on both sides. and leave an empty space between. But the muscular fibres, if you irritate them in a living body with a knife or corrosives, become shorter; their extremities approach nearer to one another, presently they are relaxed again, and these contractions and relaxations are repeated alternately for some time. The *tendons* are as void of irritability as they are of sensation. No irritation made with a knife or with any gentle corrosive can excite any convulsion in them, or produce any motion in the muscle to which the irritated tendon belongs. [680] The ligaments, *periosteum, meninges* of the brain, and all the membranes, being composed of the cellular membrane, are void of irritability. [681] The *lacteal vessels* contract and empty themselves, upon being touched with oil of vitriol, and what proves that they have a considerable share of irritability is, that though they are ever so full of chyle at the time when the animal dies, they empty themselves, and contract. [682] For by pricking it with a needle, in a dog that was almost dead, I have seen it, though not always, very frequently contract itself considerably. [683] But all the muscles are irritable. I do not know one that has not a natural palpitation after death; they all tremble, and are alternately contracted and relaxed.[685] The *æsophagus*, if it is irritated above the diaphragm, contracts itself very sensibly. [685] The *stomach* is considerably irritable, and when touched with a corrosive, its impression immediately produces a long superficial furrow in it.[686] The *intestines* both large and small, and the *cæcum* in those animals where it is large, are extremely irritable. [687] Thus by degrees I am at last got to the most irritable organ of all, viz. the heart, which as it is the cause of all motion in the human body, so is the best constructed for it, and therefore ought to be endowed with the greatest Irritability. That it is actually so, appears from experiments, especially in frigid animals, in which it is constantly irritable, and much more than the intestines. [690] From all these experiments collected together it appears, that there is nothing irritable in the animal body but the muscular fibre. And that the faculty of endeavouring to shorten itself when we touch it is proper to this fibre. From the same experiments it likewise follows, that the vital parts are the most irritable; the diaphragm frequently moves after all the other muscles have ceased, the intestines and stomach move still longer, and lastly the heart continues its motions after all the other parts are quiet.

6.7 In Sum

How an author arrives at conclusions in a historical study is especially important in works of exegesis. This chapter has identified severe problems in the use of primary and secondary texts in a 2018 historiographical monograph, focusing on suspected problems of attribution. The specific instances of suspected exposition plagiarism

6.7 In Sum

examined here are not meant be exhaustive of problems in *VFTO*, but only representative. These suspected problems impede readers who seek to understand and evaluate the claims made within the book.

In commenting on the problem of deficient attribution in a 2010 book of intellectual biography, Mark Anderson rightly notes that even though one "must of necessity adhere to an accurate chronology of events, nothing compels a specific selection of facts, quotations, or vocabulary" (Anderson 2011: 119). The same holds in books of historiography. The work of a historiographer who seeks to produce a new account of a period of intellectual history is not that of a mere compiler of undisputed facts, but evidence from primary sources must be evaluated and presented in an original way. The history of interpretation, as expressed in the secondary literature, must also be approached anew. In *VFTO*, weaknesses of originality mar the ability of the reader to weigh the merits of the study.

6.8 Postscript

In June 2018, I submitted evidence of deficient citations in two journal articles by G. to the editors of *The British Journal for the History of Science* (*BJHS*) and *SHPBBS*. I requested that the journals publish corrections for those two articles, which had been published in 2014 and 2017. At that point, a retraction had already appeared for a 2017 article by G. in *History and Philosophy of the Life Sciences* because it had been found to contain "sections that substantially overlap" with previously published work by other researchers (G. 2018a: 1). In response to my requests, *BJHS* issued an erratum the following month (G. 2018c), and a correction is forthcoming in *SHPBBS*. In publishing *VFTO*, G. had incorporated large portions of these three earlier journal articles, plus at least three others that G. had published previously from 2014–2017.

In late August 2019, however, a seven-page, paywalled online correction for *VFTO* appeared on the Springer website (G. 2019a). The correction was also inserted into the electronic version of the book, placed after the last chapter but before the bibliography. The introduction to the electronic version of the book now includes a footnote that declares that "the original version of this book was revised" and that "an erratum to this book" can be found on the publisher's website (G. 2019b: xxii).

The correction states that *VFTO* contains instances of "incomplete paraphrasis and/or lack of inverted commas" (G. 2019a: C1). What follows in the seven pages are 34 entries that provide attributions that were not present in the volume. This lengthy correction addresses a portion of the evidence of suspected plagiarism set forth in the tables of this chapter yet leaves unaddressed the principal historiographical concerns raised here.

In this 2019 correction, G. characterizes the appropriation of Mary Sunderland's encyclopedia article at the beginning of the book (see Table 6.1) as "incomplete paraphrases" (G. 2019a: C2). In a striking comment, using both underscoring and boldface type, the entry continues:

This is **the only case of lacking reference**. In all other cases, the reference is always provided in the text, footnote and/or bibliography, mostly including exact page number. (G. 2019a: C2)

This claim that no passages from any other uncredited sources appear in *VFTO* apart from Sunderland 2015 is inconsistent with the evidence discussed earlier in this chapter. For example, Table 6.5 displayed overlap with an unreferenced work by Hannah Ginsborg, and Table 6.8 had displayed overlap with an unreferenced book by Keith Peterson.

Unaddressed in the 2019 correction is the transformation of Oken's autobiographical reflections from the 1847 English translation of his work, which was displayed in Table 6.12 and analyzed in Fig. 6.2. The 2019 correction is silent on this issue, yet the re-presentation of Oken's mid-nineteenth century translated autobiographical remarks into the appearance of new third-person commentary in twenty-first century is one of the more grave uses of previously published text in the volume. Similarly, the transformation of Haller's first-person autobiographical account of his experiments into the appearance of a present-day new commentary (displayed in Table 6.13) remains unaddressed in the correction. In short, the 2019 correction is silent regarding the basic historiographical problems raised in this chapter.

In various formulations, 25 of the 34 entries of the correction—each time with underlining for apparent emphasis—state the deficiently cited source texts are mentioned somewhere in the volume. The implication seems to be that the fact that source texts were referenced in some way in the volume is somehow exculpatory for instances where attribution is lacking. For instance, the unacknowledged use of small portions of J. Smith's book on pages 17–18 of *VFTO* (displayed in Table 6.11) is addressed in the correction with the underlined observation that "The reference is provided in bibliography" (G. 2019a: C3). This implication of this underlined comment (and those like it) is that a bibliographical entry more than 100 pages later mitigates the absence of quotation marks and a proper citation. Similarly, the concession that passages from other researchers were incorporated in the book is qualified with the observation that the source works were cited elsewhere. For example, the entry that discusses the appropriation of a portion of Shirley Roe's book (displayed in Table 6.2) ends with "The reference to the work of Roe appears in footnote both in the previous and in the following page" (G. 2019a: C2). The suggestion seems to be that a reference to a source text on adjacent pages in the book serves as a kind of attribution for passages presented without quotation marks.

Accordingly, the discussion of the use of a portion of Jean-Claude Dupont's chapter (shown in Table 6.3) is qualified with the underlined sentence, "The reference is provided in footnote" (G. 2019a: C2). As noted above, Dupont's work was indeed referenced in a footnote, but only as the third of three listed entries, and not in a way that would suggest it was the source for the passage. Unaddressed also is the preservation of the use of the first-person "I" by G. that is taken from Dupont's chapter. The reader has no way of knowing that the "I" in that passage of *VFTO* is really the "I" of Dupont, rather than the "I" of G. In a similar way, the previously unacknowledged use of Miller's translation of Hegel's *Phenomenology of Spirit* (see Table 6.10) is discussed in the correction, but not the use of the first-person text where Hegel's translated words are presented as if they were G.'s words.

The concerns raised in this chapter were not simply that material is copied and pasted from sources with inadequate or no credit. If that were the principal problem with the volume, the matter could not be characterized as suspected *disguised* plagiarism. The concerns here are historiographical: the various authoritative voices are not kept distinct in key portions of the volume. The suspected plagiarism is disguised because of the blending and obfuscation of three kinds of authority: (1) the voices of the authors of primary canonical texts; (2) the voices of exegetes in the secondary literature; and (3) the voice of G. as new contributor to the research literature.

When the words of a canonical historical figure in philosophy (e.g., Kant, Hegel) or biology (e.g., Oken, Haller) are presented as the new insights by a present-day commentator, the present-day reader is misled and thereby disadvantaged in assessing their weight. The reader is led to think that she or he is hearing G., but rather the reader is unknowingly hearing Kant, Hegel, Oken, or Haller. Furthermore, when established authorities in the secondary literature summarize their own positions, and then these summaries are presented as fresh accounts of their positions by G., the reader is misled about the weight of those interpretations as well. For example, to present Lenoir's evaluation of his own position as if it were original with G. impedes the reader who is attempting to digest the history of the reception of texts.

It appears that any further corrections to *VFTO* are unlikely, as the 2019 correction boldly asserts, "this is the final correction of the reference in this volume" (G. 2019a: C1). Whether further articles by G. that were incorporated into *VFTO* will be corrected remains to be seen. The lengthy 2019 correction of *VFTO* reads more like a defense than a correction, especially with presentation of underlining and boldface font. Despite the concerns identified here, the 2019 correction does provide a measure of credit to some of the authors whose works appear in part in *VFTO* with inadequate or no attribution.

References

Anderson, Mark. 2011. Telling the same story of Nietzsche's life. *Journal of Nietzsche Studies* 42 (1): 105–120.
Bach, Thomas, and Horst Neuper. 2001. Bibliographie zu Lorenz Oken. In *Lorenz Oken (1779–1851)*, ed. Olaf Breidbach et al., 251–268. Weimar: Hermann Böhlaus.
Bertoletti, Daniele. 2018. Vital forces. *Teleology and Organization*. Verifiche 42 (1–2): 311–317.
Duchesneau, François. 2018. Foreward. In [G.], *VFTO*, v–ix. Cham: Springer.
Dupont, Jean-Claude. 2007. Pre-Kantian revival of epigenesis. In *Understanding purpose*, ed. Philippe Huneman, 37–50. Rochester: University of Rochester Press.
The embryo project encyclopedia. 2007. https://embryo.asu.edu/info/about.
[G.]. 2014. Vital forces and organization. *SHPBBS* 48: 12–20. [Correction forthcoming]
[G.]. 2018a. *Vital forces, teleology and organization*. Cham: Springer. [Corrected in G. 2019a, b].
[G.]. 2018b. Retraction note to: Diverging views of epigenesis. *History and Philosophy of the Life Sciences* 40 (2): #38, 1–2.

[G.]. 2018c. Erratum. Lorenz Oken (1779–1851): *Naturphilosophie* and the reform of natural history—erratum. *BJHS* 51 (3): 511.
[G.]. 2019a. Correction to: Chapters 1, 2, 3 and 4. In [G.], *Vital forces, teleology and organization*, C1–C7. Springer. https://doi.org/10.1007/978-3-319-65415-7_6.
[G.]. 2019b. *Vital forces, teleology and organization*. Cham: Springer. [Corrected version of G. 2018a].
Ginsborg, Hannah. 2013. Kant's aesthetics and teleology. In *The Stanford encyclopedia of philosophy* (Spring Edition), ed. Edward N. Zalta. https://plato.stanford.edu/archives/spr2013/entries/kant-aesthetics.
von Haller, Albrecht. 1936. A dissertation on the sensible and irritable parts of animals. *Bulletin of the Institute of the History of Medicine* 4 (8): 651–699.
Hegel, G. W. F. 1977. *Phenomenology of spirit*. Trans. A. V. Miller. Oxford: Oxford University Press.
Kabeshkin, Anton. 2019. Vital forces, teleology, and organization. *Philosophy in Review* 34 (2): 69–71.
Kant, Immanuel. 2004. *Metaphysical foundations of natural science*. Trans. Michael Friedman. Cambridge: Cambridge University Press.
Kanz, Kai Torsten. 2018. Vital forces, teleology, and organization. *Berichte zur Wissenschaftsgeschichte* 41 (3): 302–304.
Lenoir, Timothy. 1981. The Göttingen School and the development of transcendental *Naturphilosophie* in the Romantic Era. *Studies in the History of Biology* 5: 111–205.
Oken, Lorenz. 1847. *Elements of physiophilosophy*. Trans. Alfred Tulk. London: The Ray Society.
Peterson, Keith R. 2004. Translator's introduction. In F. W. J. Schelling, *First outline of a system of the philosophy of nature*. Trans. Keith R. Peterson, xi–xxxv. Albany: SUNY Press.
Roe, Shirley A. 1981. *Matter, life, and generation*. Cambridge: Cambridge University Press.
Smith, Justin E.H. 2011. *Divine machines*. Princeton: Princeton University Press.
Springer, n.d. Springer. https://www.springer.com/series/8916.
Steinke, Hubert. 2005. *Irritating experiments*. Amsterdam: Editions Rodopi.
Sunderland, Mary E. 2015. Abraham Trembley (1710-1784). *Embryo project encyclopedia*. https://web.archive.org/web/20160407142858/https://embryo.asu.edu/pages/abraham-trembley-1710-1784.
Toepfer, Georg. 2018. *Vital forces, teleology and organization*. NDPR, April 12, 2018. https://ndpr.nd.edu/news/vital-forces-teleology-and-organization-philosophy-of-nature-and-the-rise-of-biology-in-germany.
Weber-Wulff, Debora. 2014. *False feathers*. Heidelberg: Springer.
Zammito, John H. 2018. Vital forces, teleology, and organization. *HOPOS* 8 (2): 497–500.

Chapter 7
Template Plagiarism

Abstract In its standard variety, *template plagiarism* occurs when a plagiarist uses a previously published passage on one subject and reworks it to produce a seemingly new passage on a different subject by changing a key term. For example, altering a passage that discusses one country by substituting the name of another country produces the illusion of original research on the second country. In its more complex variety, template plagiarism may be co-extensive with another form of research misconduct—data fabrication—when the source work involves quantitative or qualitative research. This chapter considers examples of suspected template plagiarism in articles from political science and related disciplines. It begins with an analysis of the simplest variety of template plagiarism and then turns to the more complex forms that involve the appropriation of data.

Keywords Disguised plagiarism · Phrasal template · Political science · Data fabrication · Serial plagiarism · Policymaking

Template plagiarism is the use of a source text as a template to fabricate the illusion of new research. A typical case involves removing a key term from a scholarly source text and then using what remains as a template to which is added a new and often unrelated term. This procedure generates the false impression of a reliable scholarly text on a different subject. Instances of template plagiarism commonly involve the substitution of linguistic substantives of various kinds, such as proper nouns, gerunds, abstract nouns, or collective nouns. A popular version of a phrasal template is the children's game *Mad Libs*, which involves the placement of linguistic type-words within an established template to produce nonsensical yet often humorous stories. As a game, the use of templates to produce texts can be enjoyable. When this activity is used to construct fraudulent scholarly articles, however, the effects can be devastating to the quality of published research in an academic field.

Most forms of plagiarism require that the source text and the plagiarizing text concern the same subject matter. Preceding chapters have shown that this requirement typically holds in varieties of disguised plagiarism. This limitation does not generally apply, however, to cases of template plagiarism. The substitution of key terms allows a plagiarist to generate the illusion of novel research on a topic that is unrelated to the source text. The disparity in subject matter between source text and plagiarizing text in template plagiarism creates a nearly insurmountable distance between the original and the modified copy. Template plagiarism is invisible to most readers, rendering it one of the most difficult of forms of disguised plagiarism to identify.

This variety of disguised plagiarism occurs in many disciplines. (One instance from the discipline of theology was briefly considered in Sect. 5.1.1. above) Publishers can be complicit with authors of record in producing works of template plagiarism. To take a well-publicized case: in 2008 Routledge published the 168-page book *Theory for Performance Studies* in its specialized series "Theory 4." The volume was withdrawn in 2009 after researchers complained it was almost entirely constructed on the basis of Routledge's earlier 2004 volume, *Theory for Religious Studies*, another 168-page book. The principal differences between the two volumes were that in the later book, the expression "Performance Studies" had replaced "Religious Studies" and the two books had different authors of record. The publisher withdrew the later book after withering criticism (Goldingay 2009; Schechner et al. 2009). Routledge also cancelled the series, which had grown to include volumes on classics, education, and art history.

This chapter focuses on instances of template plagiarism from political science and related fields such as foreign relations, international affairs, and global studies. While plagiarism in all academic fields corrupts the downstream literature, plagiarism in these select fields produces an additional specific pernicious effect. Published research in these fields often includes recommendations for policymakers who make decisions that affect vulnerable populations around the world. When policymakers unwittingly take defective plagiarizing articles as trustworthy when forming policy, the vulnerable populations under their care can be harmed.

7.1 Standard Template Plagiarism

In 2016, the publisher Routledge re-issued a collection of essays, *The European Union Neighbourhood: Challenges and Opportunities*, which had first appeared three years earlier with the publisher Ashgate. The volume is part of a series that has a policy-making goal; volumes in the series support "ongoing networks of analysts in both academia and think-tanks as well as international agencies" (Routledge, n.d.). The editor of the collection (hereafter, "C.") also contributed a chapter titled "Threats to Human Security in the Western Balkans" (C. 2013a). It purports to be an original analysis of the effect of crime and corruption on social and economic development in the Western Balkans. Whether the chapter succeeds in its stated purpose is questionable, however, due to the prevalence of suspected template plagiarism.

7.1 Standard Template Plagiarism 129

7.1.1 A Change of Region

As its title indicates, C. 2013a ostensibly concerns security challenges in the countries of the Western Balkans. Table 7.1 presents the opening of the chapter alongside its apparent undisclosed source text, which is the beginning of an introductory chapter of an edited collection of essays on Africa published 13 years earlier, titled *Corruption and Development in Africa: Lessons from Country Case Studies* (Hope et al. 2000).

As the highlighting in Table 7.1 displays, much of the first part of the introduction of C. 2013a exhibits verbatim and near-verbatim overlap with the preface to the 2000 volume.[1] There are two significant changes, however. The term "the Western Balkan countries" replaces the term "Africa" throughout (indicated by dotted underscoring), and the longer term "organized crime and corruption" replaces "corruption" (indicated by wavy underscoring).

Even though the terms "Africa" and "Western Balkans" are collective proper nouns that refer to a defined set of countries, the terms are not synonymous either in political science, in related disciplines, or in common parlance. The term "Africa" refers to the second-largest continent, straddling the northern and southern hemispheres, covering approximately 20% of the land surface area of the planet Earth. The term "Western Balkans," however, is a contested one that can refer collectively to a portion of southern Europe consisting of Albania, Bosnia and Herzegovina, Croatia, Montenegro, North Macedonia, Serbia, and the disputed area of Kosovo in the south of Europe. The disparities of geography, ethnicity, language, culture, and other major factors make the substitution of "Western Balkans" for "Africa" inexplicable in an article that purports to analyze present-day issues in political science and international relations. The replacement of one term for the other would not seem

Table 7.1 The substitution of a region for a continent

C. 2013a: 33	Hope and Chikulo 2000: x
Introduction	Preface
In recent years, there has emerged a heightened recognition of the negative consequences of organized crime and corruption in the Western Balkan countries together with their ensuing effects on the socio-economic development process in these countries, as well as their corrosive impact on society and on the fledgling democratization process. The two phenomena are linked to the climate of unethical leadership and bad governance. Increasingly, the issue has come to the fore as a concern expressed by Western countries and international organizations such as the European Union (EU). As a consequence, not only have the two issues been moved to the top of the policy reform agenda, but a high premium has also been placed on combating them. Controlling and eradicating organized crime and corruption has therefore taken on even greater significance in the stabilization and democratization process of Western Balkan countries. Organized crime and corruption negatively affect the development process at administrative, economic, political and social levels, threatening human security in the Western Balkans.	In recent years, there has emerged a heightened recognition of the negative impact of the pandemic of corruption in Africa and its ensuing negative consequences on the socio-economic development process, as well as its corrosive impact on society and on the fledgling democratization process. Corruption is shown to be linked to the climate of unethical leadership and bad governance [...].Increasingly, the issue has come to the fore as a problem of concern expressed by Western donor countries and international organizations. Consequently, not only has corruption been moved to the top of the policy reform agenda but a high premium has also been placed on combating it. Controlling and eradicating corruption has, therefore, taken on even greater significance in the quest for development in Africa. [...] Corruption negatively affects the development process at the administrative, economic, political and social levels.

---

[1] The apparent defects of C. 2013a were examined as a research project I conducted with students enrolled in the course "Critical Research and Writing" at Ohio Dominican University during the Spring 2019 semester. We documented our findings and sent a retraction request to the publisher on 4 April 2019.

likely to produce a document reaching a level of precision required for useful policy recommendations.

Both the source text and the apparently derivative plagiarizing text concern the effects of corruption on socio-economic development in fledgling democracies, yet this shared theme does not render the apparent template plagiarism to be innocuous. Furthermore, the striking overlap does not seem to be accidental or the result of chance. There is public evidence that C. is familiar with the undisclosed source text, since C. briefly cites with quotation marks a different part the book elsewhere (C. 2013a: 36). Yet no quotation marks or citations are present to indicate to the reader that this introductory portion of the chapter on the Western Balkans overlaps with the previously published introductory portion of the book on Africa.

Template plagiarism is a form of *disguised* plagiarism because the substitution of key terms in the source text with unrelated ones in the plagiarizing text creates additional distance between the two. In this case of suspected template plagiarism, readers are unlikely to recognize that a passage that ostensibly deals with the Western Balkans overlaps with another one dealing with the continent of Africa published years earlier by researchers with expertise in the particularized political and economic conditions of the continent. As a disguised form of plagiarism, template plagiarism is especially difficult to identify because the substitutions of key terms causes an illusion of unrelatedness between the source text and plagiarizing text.

7.1.2 A Change in Topic

Other portions of C. 2013a also appear to exhibit template plagiarism where major terms have been removed from an undisclosed source text, replaced with new ones, and then the altered passage is set forth as a novel analysis. Table 7.2 exhibits a passage purporting to be a consideration of "organized crime and corruption" in the Western Balkans but it apparently derives from a book on illegal immigration. Since "corruption" and "illegal immigration" are not synonymous, the overlap of the earlier source text and the later text is disguised, and the quality of the analysis in the later derivative text must be judged as questionable.

The passages are almost identical save three differences. First, the subject term "corruption" replaces "illegal immigration." Second, the more specific term "Western

Table 7.2 The substitution of a subject term

C. 2013a: 34	Mitsilegas et al. 2003: 42
Organized crime and corruption have come to be seen as major problems in Western Balkan countries. These are by nature transnational problems as they occur across recognized frontiers and yet do not comprise state-sponsored activities. Organized crime and corruption are perceived as security problems that need to be countered by a range of policies and have become linked together in the minds of policymakers.	Organised crime and illegal immigration have come to be seen as major problems in Europe [...]. These are by nature transnational problems as they occur across recognised frontiers and yet do not comprise state-sponsored activities. Organised crime and illegal immigration are perceived as security problems that need to be countered by a range of policies and have become linked together in the minds of policy-makers [...].

7.1 Standard Template Plagiarism

Table 7.3 A change in references to secondary literature

C. 2013a: 37	Shani 2007: 4
During the 1990s, an alternative approach emerged that sought to reconceptualize security by making the individual human being and not the state the main referent object of security (Tadjbakhsh 2005). This came to be termed the 'human security' approach (ibid.). It is suggested here that this approach, rather than the national security paradigm, may enable us to better analyse how society in Western Balkan countries is being affected by organized crime and corruption.	During the 1990s an alternative approach emerged which sought to reconceptualize security by making the individual human being and not the state the main referent object of security.[7] [[7] Buzan et al. 1998: 36] This came to be termed the 'human security' approach or agenda.[8] [[8] Held 2004] It is suggested here that this approach rather than the national security paradigm may enable us to better respond to the principal sources of insecurity in the post 9/11 world.

Balkan countries" replaces the more general term "Europe." Finally, the orthography of two terms has been changed from British English to American English as "s" becomes "z" in the terms "organized" and "recognized." There is no reference anywhere in C. 2013a to the undisclosed source text, either a through footnote, in-text citation, or bibliographical entry. Readers of C. 2013a have no way of knowing from the text itself that they are encountering a passage—albeit in modified form—that overlaps with a work published a decade earlier on a different subject.

The passages of C. 2013a exhibited in Tables 7.1 and 7.2 are not outliers of the chapter, but rather are representative of the apparent deficiencies of the scholarship that mar the reliability of the publication. Many other passages in C. 2013a appear to be constructed by using undisclosed source texts as templates (and nearly every page of the article is marred by apparent copy-and-paste plagiarism). The switching of key terms is not always as substantial as the replacement of a continent for a small European region. More typically, the addition of qualifiers particularize general claims found in an apparent source text. Table 7.3 shows how broad claims about responding to insecurity "in the post 9/11 world" appear to have been reworked in C. 2013a to pertain to crime and corruption in Western Balkan countries.

In this instance, the suspected template plagiarism involves the substitution of terms as well as the substitution of references to the secondary literature, and this procedure creates further distance between the source text and the apparently derivative text. Despite these changes, the role of the suspected source text as a template for the passage is still evident through a side-by-side comparison. The expression "It is suggested here" is repeated in C. 2013a, but of course the "here" cannot retain its original meaning when it is shows up in the newer context.

Table 7.4 presents another example where a general claim in an undisclosed source text appears to have been shochorned into a new discussion. A discussion of anti-corruption measures in the example countries of "Italy, Japan, Belgium" has been modified to so that the examples are "Croatia or Macedonia." In this way, the source

Table 7.4 A change in example countries

C. 2013a: 45	Bull and Newell: 2003: 244
There have been an increasing number of anti-corruption actions and measures, and in some countries, such as Croatia or Macedonia, this has led to a number of high-profile arrests and convictions.	There have been an increasing number of anti-corruption actions and measures, and in some countries (e.g., Italy, Japan, Belgium) this has led to a number of high profile arrests and convictions.

text is altered so that the apparently plagiarizing version relates more particularly to the Western Balkan region.

The use of an already published text as a template, with the addition of new key terms and different citations, does not justify the foregoing of a reference to the source text or the abandonment of quotation marks around the verbatim portions that derive from the source text. Acts of template plagiarism violate the accepted conventions of attribution for scholarly writing. In the examples presented in Tables 7.2, 7.3 and 7.4, the apparent source texts by Mitsilegas et al. (2003), Shani (2007), and Bull and Newell (2003) are nowhere referenced in C. 2013a.

It is difficult to imagine that articles constructed using template plagiarism could be meaningful and valuable contributions to an academic discipline. The presumed rigor proper to a discipline, and the specificity of truth claims, would seem to rule out the use of template plagiarism as a method for constructing research works that possess utility to the community of scholars. Furthermore, the latent presence of many published articles based on template plagiarism in a given discipline would be evidence of a widespread institutional inability to defend against this particular form of research misconduct. If an author of record has enjoyed continued success in placing the products of template plagiarism in established journals in books by reputable publishers over many years, so that these plagiarizing counterfeit articles are taken as genuine such by practitioners of the discipline, the threat to research integrity must be judged as substantial.

7.1.3 Changes in Region and Topic

As noted above, the standard form of template plagiarism involves the substitution of key terms—usually substantives—to produce the illusion of new research. Table 7.5 presents an example of suspected template plagiarism from the beginning of an article on Serbia by C. published in 2011. Titled "Human Rights Promotion in Serbia: A

Table 7.5 A change in country and in topic

C. 2011: 142	Reinhard 2008: 3–4
Introduction	Introduction
Accession to the European Union (EU) is oftentimes considered as the most successful instrument for the promotion of human rights in post-communist countries, such as the Western Balkans. As the democratisation of non-member states is both a normative and strategic aim of the EU, human rights promotion is a main element of its foreign policy. It is reflected in its relation with these countries in general, and in the enlargement policy, in particular. Even though the membership perspective might be a promising instrument to promote democracy, and human rights in external countries, the underlying causal mechanisms have to be identified in order to provide evidence for this assumed causality. Conditionality serves in this context both as a promising tool of the EU to promote democracy and human rights and as a theoretical framework to explain causalities between an EU membership perspective and the implementation of human rights values in Serbia.	Accession to the European Union (EU) is oftentimes considered as the most successful instrument for the promotion of democracy in post-communist countries. […] As the democratisation of non-member states is both a normative and strategic aim of the EU, democracy promotion is a main element of its foreign policy. It is reflected in its relation with third countries in general […]. Even though the membership perspective might be a promising instrument to promote democracy in external countries, the underlying causal mechanisms have to be identified in order to provide evidence for this assumed causality. Conditionality serves in this context both as a promising tool of the EU to promote democracy and a theoretical framework to explain causalities between an EU membership perspective and successful democratisation process in the target country (Kubicek 2003; Kneuer 2007; Pridham 1997; Schimmelfennig/Sedelmaier 2005).

Difficult Task for the European Union," the article was published in *Revista Brasileira de Política Internacional* (*RBPI*). The apparent source text, however, is not a study of Serbia, but rather is a study of the Ukraine. With the substitution of a few fundamental terms, a consideration of "democracy in the Ukraine" is reincarnated as a consideration of "human rights in Serbia."

The apparent source text of the 2011 passage is the beginning of a 2008 article by Janine Reinhard that analyzes how the European Union uses the mechanism of conditionality to promote democracy in that region. As the comparison of the two versions shows, the heading, the order of the sentences, and the manner of expression overlap for the most part. The main alteration is that the term "human rights" replaces "democracy" in the later text, and the terms "the Western Balkans" and "Serbia" are added to the first and last sentences. There is no reference to Reinhard 2008 in the introduction to C. 2011, but a later portion of the article contains a brief reference in a discussion of an unrelated point (C. 2011: 144). The five references to previously published works credited at the end of the passage in Reinhard's original are not repeated in C. 2011's version of the passage, presumably because the references concern democratization (the subject of Reinhard 2008) more than human rights (the ostensible subject of C. 2011).

The article in *RBPI* has been well received in the downstream literature. In addition to self-citations to it by C. (e.g., C. 2016: 118), other researchers have engaged it with positive citations in their articles (e.g., Uğurlu 2013: 180). These researchers seem to believe that C. 2011 is a reliable analysis of the European Union's role in supporting human rights in Serbia.

7.1.4 A Change in Genre

In addition to changes in region, country, or topic, sometimes the nature of the source text requires a further modification of a plagiarizing text in cases of template plagiarism. Two early pages of a 2013 article in *European Foreign Affairs Review* by C. substantially overlap with pages from the early part of unreferenced 2009 thesis by Petya Mandazhieva from Central European University (C. 2013b: 430; Mandazhiev 2009: 1–2; C. 2013b: 432–433; Mandazhieva 2009: 33, 36). A key change in the derivative text, beyond a substitution of the words "in the Western Balkans" for "European," is the replacement of Mandazhieva's expression "of this dissertation" with the new expression "in this article" (C. 2013b: 430; Mandazhieva 2009: 2). Had the original words "in this dissertation" been preserved, it would be clear to readers that the passages in 2013 article have already appeared in another format—the unpublished master's thesis—prior to their appearance as a journal article.

7.2 Template Plagiarism Versus Template Text Recycling

Sometimes authors re-use their previously published texts without acknowledgment in later publications. This contested practice of text recycling (sometimes called "self-plagiarism" and "inappropriate text reuse") admits of degrees of gravity (Horbach and Halffman 2019; Pemberton et al. 2019). In its extreme forms, the unacknowledged reuse of a complete work constitutes an act of duplicate or redundant publication, and such cases typically warrant retraction by editors and publishers (Dougherty 2018: 75–80). Some researchers produce works by using their own previous works as templates for later ones, creating the phenomenon of what can be called *template text recycling* (or "template *self*-plagiarism"). Much like standard form of template plagiarism, this related phenomenon of template text recycling creates the appearance of a new, original work. Nevertheless, the two should be distinguished as separate phenomena. Template plagiarism denies the genuine author credit for composition, but template text recycling does not.

Table 7.6 displays passages from three articles that share the theme of post-communist European integration. The overlap of the passages is substantial, even though professedly the first article concerns Poland (C. et al. 2007), the second Croatia (C. 2009), and the third Albania (J. 2013). C. is a common author of record to the first two articles, but not the third. All three articles appear to share the same basic template for their respective introductions, but the key term in each (i.e., "Poland," "Croatia," "Albania") is different. The earliest is from a book chapter from an edited collection on globalization published by Elsevier; the second from the Portuguese political science journal *Nação e Defesa*, and the last from the published proceedings of an international conference on the topic of democracy.

Despite the overlap common to these passages, their precise relationships to each other are not clear. For example, one of the following situations could be the case:

Table 7.6 A template in passages on Poland, Croatia, and Albania

C. et al. 2007	C. 2009	J. 2013: 76
This essay aims to address the simultaneously interconnected and heterogeneous responses of Polish post-communist course of change to global and regional processes, including European integration. In this line of research, we search for answers to how the linkages among globalization, regionalization, and Europeanization work in the case of Polish post-communist transition. This will be pursued through an analysis of the democratization course, mainly regarding political, institutional and social aspects, and economic integration. Despite elements of complementarity and resistance in the working relationships among the three concepts, which are highly debatable, we find they have substantial implications on Polish policymaking.	This essay aims to address the simultaneously interconnected and heterogeneous responses of Croatia's post-communist course of change to regional process of European integration. In this line of research, we search for answers to how Europeanization works in the case of Croatia's post-communist transition. This will be pursued through an analysis of the democratization course, mainly regarding political, institutional and social aspects, and economic integration. Despite elements of complementarity and resistance in the working relationship among the EU and Croatia's, which are highly debatable, we find they have substantial implications on Croatia policy-making.	This paper aims to address the simultaneously interconnected and heterogeneous responses of Albania's post-communist course of change to European integration. In this line of research, we search for answers to how Europeanization affected Albania's post-communist transition. This will be pursued through an analysis of the democratization course, mainly regarding political, institutional and social aspects, and economic integration. Despite elements of complementarity and resistance in the working relationship among Europeanization and Albanian government policies, which are highly debatable, we find it has substantial implications on Albanian policy-making.

(1) All three passages (C. et al. 2007; C. 2009; J. 2013) are based on an earlier, undisclosed, and unknown template
(2) The earliest passage (C. et al. 2007) is the template for the later two (C. 2009; J. 2013)
(3) The earliest passage (C. et al. 2007) is the template for the middle passage (C. 2009), which in turn is the template for the latest passage (J. 2013).

If situation (1) is the case, all three passages are cases of suspected template *plagiarism*. If instead, either situation (2) or (3) is the case, then the earliest passage is authentic, the middle passage is the product of suspected template *text recycling*, and the latest passage is the product suspected template *plagiarism*.

A reader of the middle or latest passage is unlikely to suspect at the time that she or he is reading a text that could be the product of a template. That is, most readers will likely remain unaware that the declared subject matter in the introduction of each article—Croatia or Albania—is simply the result of a different principal substantive having been inserted into a template in the fabrication of the texts. The use of a template by distinct author groups in different venues suggests an almost mechanical approach to the generation of some scholarship. In light of the many differences—cultural, economic, political, linguistic, and societal—of Poland, Croatia, and Albania, one must wonder about the quality of scholarship on those countries that is dependent on templates. The use of the specific template common to C. et al. 2007, C. 2011, and J. 2013 does not appear unique. A single-authored chapter by C. in a 2014 collection by Lexington Books also seems to use, but to a lesser degree, the same template in its introduction (C. 2014a: 246–247).

These template-based articles continue to influence the downstream scholarly literature. Researchers across fields appear to treating C. et al. 2007 as reliable; journals in diverse areas such as supply chain management and political methodologies have articles that discuss it (e.g., Abylaev et al. 2014: 58; Lavdas 2017: 109). C. 2009 is extracted and discussed in a doctoral dissertation (Stojanova 2013: 76) and in a monograph (Dallara 2014: 41–42), and is cited in conference proceedings and journal articles. None of these researchers who engage C. et al. 2007 and C. 2009 raise any questions about the quality of the research presented in them.

7.3 Template Plagiarism Involving Data Fabrication

The examples of suspected template plagiarism considered above involve the removal of terms from a source text and their replacement with new ones, thereby creating the appearance of original analyses on a new subject. They exhibit suspected *standard* template plagiarism. A more complex variety occurs when the source text contains data, so that the data set from the original context is reasserted in the new, plagiarizing version.

The preceding examples of suspected standard template plagiarism involved either passages from the introductions of the articles (Tables 7.1, 7.5 and 7.6) or relatively short passages that might be considered by some readers to be ancillary to the main portion of an article (Tables 7.2, 7.3 and 7.4). Some researchers judge academic plagiarism to be significant only if it concerns the portions of articles that constitute the "core" of a given article (see St. Onge 1988; Bouville 2008). On this line of reasoning, claims of suspected plagiarism should be reserved to cases in which there is unacknowledged duplication in portions of articles that present data, disclose new research findings, or offer novel conclusions. On that view, the other portions of articles, such as introductions, literature reviews, method statements, bibliographies, and references, should not be treated with the same expectations of originality, and some overlap should be judged to be allowable in those purportedly non-essential portions of an article (see Roig 2015:14; Chaddah 2014:127). Although this view has been criticized (Dougherty 2018: 63–67), it has its defenders. The complex forms of template plagiarism, involving the reporting of data, deserve consideration.

7.3.1 Fieldwork Interviews and Template Plagiarism

In some branches of political science and related disciplines, research includes qualitative data obtained through fieldwork interviews. In 2013, C. published an article in the Taylor & Francis quarterly *Journal of Contemporary European Studies* that presents qualitative fieldwork data. Titled "Civil Society in Macedonia's Democratization Process," the article claims to be based in part on fieldwork "conducted by the author with a representative of a CSO [Civil Society Organization] in Macedonia. Skopje, February 2011" (C. 2013c: 211, n. 6 and n. 7). Two explicit quotations from the anonymized interview are cited in the article. The difficulty in trusting the reliability of the reported fieldwork is that the quotations ascribed to the anonymized Macedonian interviewee in the article correspond verbatim and near-verbatim to a pair of passages from an EU document. Furthermore, the sentences in the article that introduce and comment upon the putative words of the anonymized Macedonian interviewee also overlap with sentences in the EU document. In short, the EU document seems to be the undisclosed source text for the purported fieldwork quotations and the analysis that surrounds them. The overlap calls into question the veracity of the reported fieldwork and also the conclusions of the 2013 article.

Table 7.7 presents a passage from C. 2013c containing first putative fieldwork quotation alongside the suspected source text, the 2011 EU report. Highlighting indicates the overlap between the two, and underlining marks the words common to the putative fieldwork quotation and the parallel words in suspected source text.

In C. 2013c, a sentence of 18 words is explicitly ascribed to the Macedonian CSO representative, both in both the body of the text and through a footnote. This same sentence—a string of 18 consecutive words—is found verbatim in the EU report. Yet in the EU report, the 18 consecutive words are not represented as part of a fieldwork interview but simply appear as scholarly analysis written by the authors of that report. The highlighting shows that the texts that come both before and after the putative fieldwork quotation in C. 2013c also happen to overlap with the same EU document.

7.3 Template Plagiarism Involving Data Fabrication   137

Table 7.7 A Macedonian fieldwork interview statement concerning CSO funding

C. 2013c: 211	TACSO 2011: 10–11
In 2007 a Code of Good Practices for the financial support by government of citizens' associations and foundations was adopted. This set basic criteria that were to be fulfilled by CSOs in order to receive state funding. Nevertheless, since this is not an obligatory act for state bodies, government institutions rarely allocate support to CSOs in a transparent manner according to clear and equitable criteria. Very often funds are allocated to arbitrarily pre-selected beneficiary organizations, and only a very few state institutions distribute funds through open calls to tender. As a representative of a Macedonia CSO stated: '<u>What is still missing in the process is monitoring of the project's implementation and evaluation of the results.</u>'[7] In addition, many CSOs still claim that political parties and official bodies affiliated to CSOs have significant influence on the decisions.	In 2007 a Code of Good Practices for the financial support by government of citizens associations and foundations was adopted and set following: basic criteria that should be fulfilled by CSOs in order to receive state funding […]. However, Code of Good Practices is not obligatory act for state bodies, thus government institutions rarely allocate support to CSOs in a transparent manner according to clear and equitable criteria. Very often funds are allocated to arbitrarily pre-selected beneficiary organisations and only a very few state institutions distribute funds through open calls to tender. […] However, <u>what is still missing in the process is monitoring of the projects implementation and evaluation of the results.</u> In addition, many CSOs still claim that political parties and affiliation of CSOs with the authorities have significant influence to the decisions.
[7] Interview conducted by the author with a representative of a CSO in Macedonia. Skopje, February 2011.	

There are no quotation marks or citations to indicate to the reader that the whole passage has already substantially appeared in print two years earlier in a different context.

It appears that C. has copied a passage from the EU report, fashioning one of the sentences into a fieldwork quotation. This procedure requires the addition of various disguising elements to what was originally a simple sentence of scholarly analysis. The additions include quotation marks, a new speaker (a Macedonian CSO representative), a location (Skopje), and a date (February 2011).

This passage from C. 2013c is a case of suspected *template* plagiarism because the previously published EU report is apparently being used as a template for the manufacture of qualitative fieldwork data. If it were not for the apparent fabrication of the fieldwork quotation, the passage from C. 2013c could be classified as a simple suspected case of copy-and-paste plagiarism, not an instance of disguised plagiarism. Plagiarism of data is a form of data fabrication, since only in its original context is the data real. The addition of the abovementioned circumstances (such as the speaker, location, and date) transform a sentence of scholarly analysis into a presentation of qualitative fieldwork data. Readers of C. 2013c are unlikely to realize that the words they encounter as being ascribed to the anonymized Macedonian interviewee, represented to have been uttered to the author of record of C. 2013c during on-the-ground fieldwork in 2011, appear to have been forged from the EU report, which does not present the words as qualitative fieldwork data. That is, what appears to readers as a firsthand account transmitted by the author of record obtained during a research fieldwork interview is really the scholarly analysis of a political situation by EU employees.

As noted above, this instance of suspected disguised plagiarism is not an outlier; the same parameters attend the second fieldwork quotation set forth in C. 2013c. Table 7.8 presents it alongside the apparent undisclosed source text. Again, a sentence of scholarly analysis in a previously published work has been repackaged for readers as a new, unique statement conveyed in confidence by an anonymized interviewee in Macedonia.

138   7   Template Plagiarism

Table 7.8 A Macedonian fieldwork interview statement on social progress

C. 2013c: 211	TACSO 2011: 15
However, as a representative of a CSO in Macedonia stated: 'There has not been any progress in establishing effective dialogue with civil society regarding policy making particularly in the preparation of the state budget, or in improving financial support for CSOs from public funds.'[6] Without sufficient autonomy, the Unit cannot take proactive measures towards implementing the strategy or in establishing direct communication with civil society (MCIC 2011, 13). The implementation of the strategy is perceived to proceed more quickly when assisted by external finance or when an action is linked to the process of European integration—either in connection with the establishment of European standards or the convergence of Macedonian law to the *Acquis Communautaire*. [6] Interview conducted by the author with a representative of a CSO in Macedonia. Skopje, February 2011.	Areas where no advancement was achieved are: establishing effective dialogue with civil society and CSOs' participation in policy making, particularly in the preparation of the state budget; improving the financial support of CSOs from public funds and development of CSOs in rural areas. Implementation of the Strategy is perceived to proceed quicker when assisted by external finance or when an action is linked to the process of European integration – either in connection with the establishment of European standards or the convergence of Macedonian law to the *Acquis Communautaire*.

In both the body of the text of C. 2013c and in an accompanying footnote, the putative fieldwork quotation is presented as a documented word-for-word statement from an anonymized CSO representative. The words that compose it, however, largely correspond to a portion of the EU report with only minor differences. Furthermore, a significant portion of the text that follows the putative fieldwork interview quotation also overlaps with the EU report. The verbatim and near-verbatim overlap of both (1) the text of the putative fieldwork quotation and (2) the surrounding sentences with the EU report suggests that either the fieldwork interview that forms the basis of C. 2013c either did not occur at all or did not occur as described. On either account, the reliability of the article is vitiated, since it appears that C. is using previously published research as a template to fabricate qualitative research data.

In January 2019, the *Journal of Contemporary European Studies* published a 10-page "Correction" to C. 2013c, (C. 2019a). The length of the correction is remarkable, since it is over half as long as the original 16-page article. The correction begins by stating that when C. 2013c "was published online and in print, the following punctuation and citation were missed" (C. 2019a: iv). What follows in the correction are portions of the original article, modified with the addition of quotation marks as well as citations to previously uncredited source texts. Despite the extensive addition of quotation marks and citations to the article in the correction, the passages set forth in Tables 7.7 and 7.8 are principally unaffected by the 2019 correction. There is a slight modification made to the passage containing the first fieldwork quotation displayed in Table 7.7; the initial part that states, "What is still missing in the process is monitoring" has been updated to "it still lacks the monitorization" (C. 2019a: ix). The first five words have been altered, but the rest remains the same, including the surrounding words that are apparently plagiarized. The passage displayed in 7.8 remains unchanged by the 2019 correction. Despite changes to the other portions of C. 2013c by the addition of quotation marks and citations, the 2019 correction still apparently vouches for the integrity of the fieldwork interview quotations. That is, the qualitative data set in the 2013 article is still represented as being trustworthy, despite the rectification of other failures to attribute sources correctly.

7.3.2 Evaluating Evidence of Template Plagiarism Involving Fieldwork

The apparent unreliability of the qualitative data extends beyond C. 2011. In 2016, C. published an article in *Journal of Contemporary Central and Eastern Europe* titled "Bridging the Gap: The Serbian Struggle for Good Governance." The article claims to be based on anonymized fieldwork interviews "carried out with some representatives of non-governmental organizations in Belgrade in March 2014" (C. 2016: 114). The putative interviewees also include "a Serbian judge" in addition to the NGO representatives (C. 2016: 124). The difficulty in accepting the reliability of the qualitative data presented in the 2016 article is that the interview quotations set forth in the article corresponds verbatim and near-verbatim to previously published sentences found in the earlier research literature by others. Furthermore, in some cases the texts that surround the putative interview quotations also are present in the various suspected source texts. It seems that prior published research by others is being used as a template to manufacture qualitative research data.

Table 7.9 presents the first interview quotation in C. 2016 ascribed to fieldwork conducted by the author of record in Belgrade in March 2014. The main impediment to accepting its reliability is that 24 of the 26 words correspond to a portion of a December 4, 2013 newspaper article by the Serbian press organization RTV B92.

The words ascribed in C. 2016 to an anonymized NGO representative are the words that appear in a the Serbian newspaper story in December 2013, three months prior to C.'s declared interview date of March 2014. In the newspaper story, the speaker of the words is identified by name as Vladimir Goati, president of Transparency International Serbia. What C. 2016 presents as an anonymized interview quote overlaps with a public statement by a public figure. Furthermore, as presented as the newspaper story, Goati's words are not offered not in the manner of a verbatim quotation, but rather as a journalist's summary of the main points of Goati's views about transparency in Serbia.

In the apparent manufacture of the 2014 fieldwork quotation, C. has apparently used the 2013 newspaper article as a template. It appears that the quotation has been manufactured by concealing Goati's name and his organizational affiliation. Then, the journalist's summary of Goati's position is forged into direct quotation, with the addition of quotation marks, an attribution to an anonymous NGO representative, a 2014 date, and a general location (Belgrade).

Table 7.9 A Serbian fieldwork interview on anti-corruption laws

C. 2016: 122	RTV B92 2013
As the representative of a non-governmental organization in Serbia argued, some of Serbia's main problems are "the violation of anti-corruption laws, the lack of sufficient capacities on the part of the supervisory bodies charged with their implementation, and insufficiently transparent decision-making processes".[8]	The Transparency International representative from Serbia said that some of Serbia's biggest problems were violations of anti-corruption laws, the lack of sufficient capacities on the part of the supervisory bodies charged with their implementation and insufficiently transparent decision-making processes.
[8] Interview conducted with a representative of a non-governmental organization in Belgrade (March 2014).	

Table 7.10 A Serbian fieldwork interview on criminal influences

C. 2016: 124	Ağir 2009: 5
The weak structures of the state encourage the threat of criminal activities, due to the "ease with which criminal organisations are able to penetrate the state and its institutions".[12]	The structures of weak states encourage the threat of criminal activities and their profitability, due to the ease with which criminal organizations are able to penetrate the state and its institutions.
[12] Interview conducted with a representative of a non-governmental organization in Belgrade (March 2014).	

A second putative fieldwork interview quote presented in C. 2016 is set forth in context in Table 7.10. This time the apparent source text appears to be simply a sentence of scholarly analysis from a 2009 journal article on security titled "Rethinking Security in the Balkans" by Bülent Sarper Ağir. The 14 words quoted in C. 2016 and ascribed to "a representative of a non-governmental representative in Belgrade (March 2014)" correspond verbatim to a part of a sentence in Ağir's article. Furthermore, the text that precedes the putative fieldwork quotation also overlaps to a significant degree with the same sentence in Ağir's article. In the original setting in the 2009 article, the words are not presented as any kind of quotation from an NGO official but simply are part of Ağir's critical scholarly analysis of criminal activity in the Balkans.

The overlap of the putative fieldwork quotation with suspected source text is verbatim, except that the orthography of the word "organization" has been changed from American English to British English.

To accept this fieldwork interview quotation as reliable, one must accept as true the following entirely implausible scenario: *In March 2014, a NGO representative shielded with research anonymity responds to questions on the sensitive subject of regime corruption by uttering words to an interviewer that correspond verbatim and near-verbatim with a 2009 journal article published half a decade earlier. Furthermore, when the author of record of C. 2015 writes up the account of the interview two years later, the author of record introduces the fieldwork quotation using words that also happens to line up verbatim and near-verbatim from the same sentence of the 2009 journal article.* It seems unreasonable to assume that the 2009 article would predict with accuracy the words of an interviewee five years before they are uttered, as well as predict with accuracy the words of the interviewer seven years before they are written, and both predictions would occur in the same sentence on one page in a relatively obscure Turkish academic journal. It seems more reasonable to assume that the journal article is being used as template for the production of qualitative research data.

The putative interviews in Serbia are asserted to have taken place in March 2014 by the author of record for C. 2016. The date is again relevant for evaluating the veracity of the fieldwork quotations. For the third quotation in the article, the apparent source text is a November 2006 conference presentation titled "Organized Crime in the Western Balkans" by Dejan Anastasijevic, a senior investigative reporter from the Serbian weekly *Vreme*. Table 7.11 presents the putative fieldwork quotation in context alongside the corresponding passage from Anastasijevic's paper.

7.3 Template Plagiarism Involving Data Fabrication

Table 7.11 A Serbian fieldwork interview on law enforcement

C. 2016: 124	Anastasijevic 2006: 3
Although there are efforts to improve the legislation, most of them are instigated from outside, rather than by local governments' awareness of the social and economic cost inflicted by the activities of organized criminals. According to a Serbian judge, "the country needs <u>law enforcement agencies, and</u> needs to <u>create regional networks for fighting organised crime</u>".[14] [14] Interview conducted with a Serbian judge in Belgrade (March 2014)	Still, the situation is not totally bleak. In recent years, most countries in the region have made efforts, in various degrees, to improve their legislation, reform the <u>law enforcement agencies, and create regional networks for fighting organized crime</u>. However, most of these efforts are instigated from the outside, rather then by local governments' awareness of the social and economic cost inflicted by the activities of organized criminals.

The putative fieldwork quotation is said to be result of a March 2014 interview with "a Serbian judge in Belgrade." The quotation and the surrounding text overlap substantially with consecutive sentences from Anastasijevic's paper. This time the veridicality of the words ascribed to the judge require that the judge uttered words in 2014 that correspond to a sentence of scholarly analysis that Anastasijevic wrote for the First Annual Conference on Human Security, Terrorism and Organized Crime in the Western Balkan Region. Furthermore, the reliability of the quotation also requires the implausible scenario that the author of record of C. 2016 is unknowingly using verbatim and near-verbatim words by Anastasijevic in his conference presentation in introducing the fieldwork quotation, words that occur on the same page of the presentation paper. The principle of simplicity suggests that a more likely scenario is that C. 2016 is using the text of the 2006 conference presentation as a template in the in manufacture of the qualitative fieldwork data.

An additional putative fieldwork quotation in C. 2016 is professed to have been uttered to C. by a "representative of a non-governmental organization," but it overlaps verbatim with an unreferenced sentence in a 2012 essay by corporate consultant Nalin Jayasuriya in the Sri Lankan newspaper *FT Daily*. In the original setting in the 2012 article, the words are not presented as a quotation from an NGO official but are simply are part of Jayasuriya's original analysis of government corruption in the Balkan region. As Table 7.12 displays, however, Jayasuriya's essay appears to have been appropriated not only in C. 2016 in the manufacture of a fieldwork quotation, but also earlier in C. 2013a, which is the 2013 book chapter published by Ashgate/Routledge discussed above.

Table 7.12 Suspected template plagiarism versus suspected copy-and-paste plagiarism

C. 2013a: 50	C. 2016: 124	Jayasuriya 2012
Corruption in the region remains widespread in all sectors, from the healthcare system to customs and tax institutions and parliament. This situation results from weak laws, <u>inadequate investigations by police, corrupt judges and politicians, insufficient sentencing, and a lack of coordination of anti-corruption efforts</u>.	Corruption remains widespread in all sectors. This situation results from "<u>inadequate investigations by police, corrupt judges and politicians, insufficient sentencing and a lack of coordination of anti-corruption efforts</u>".[13] [13] Interview conducted with a representative of a non-governmental organization in Belgrade (March 2014).	Corruption in the region remains widespread in all sectors, from the healthcare system to customs and tax institutions and the parliament. The reasons are myriad: weak laws, <u>inadequate investigations by police, corrupt judges and politicians, insufficient sentencing and a lack of coordination of anti-corruption efforts</u>.

Table 7.12 demonstrates how the same source text can be appropriated in different kinds of suspected plagiarism. When a portion of Jayasuriya's 2012 essay appears without attribution in C. 2013a, one finds merely a case of suspected copy-and-paste plagiarism. In C. 2013a the suspected source text is not being modified significantly, but is simply being copied with minor alterations and without attribution. The situation is much different when a portion of Jayasuriya's 2012 essay appears without attribution in C. 2016, since to the words have been transformed through the addition a speaker (an anonymized NGO representative interviewee), a time in which they were purportedly uttered (March 2014), a location (Belgrade), and a context (an anonymized fieldwork interview). The undisclosed suspected source text—Jayasuriya's 2012 essay—is being used, apparently, as a template to which significant changes are being made for the manufacture of qualitative research data. Again, in this instance, it is crucially important to note that the words that introduce the purported fieldwork quotation, as well as the words of the purported fieldwork quotation, both appear together in the suspected source text.

It is an oddity that C. appears to be using the suspected source text in two ways in the two articles. The same words that C. presents as original analysis in 2013c are later presented as qualitative data from an anonymized interview in 2016. Even if Jayasuriya's (2012)'s essay had not been discovered as the suspected urtext for both articles, so that one only possessed the text of C. 2013a, 2016, there would be sufficient evidence to question the quality of the data in C. 2016. The reason is clear: the words presented as original to the author of record in C. 2013a are re-presented by three years later by C. as the words of an anonymized NGO representative.

To assume that C. 2013a, 2016 are reliable articles, one has to assume that three individuals on four different occasions have unwittingly expressed the same words in the span of five years:

(1} [2012] A Sri Lankan corporate consultant writing for *FT Daily* (Jayasuriya)
(2) [2013] A researcher doing original analysis in a book chapter (C.)
(3) [2014] A Serbian NGO representative giving a statement (anonymous interviewee)
(4) [2016] A researcher introducing of the words of the NGO representative in a journal article (C.)

This scenario is extremely unlikely. It is more reasonable to assume that Jayasuriya's (2012) *FT Daily* essay is the urtext for a copy-and-paste plagiarism in C. 2013a and template plagiarism in C. 2016.

Positive citations to C.'s articles continue, and C. 2016 is no exception. It is subject to self-citation in later works (e.g., C. 2019c: 161) and journal articles in various European languages. The reduction of all of the qualitative data presented in the article to previously published and undisclosed source articles suggests that C. 2016 is not reliable and that its conclusions—based as they are upon the data—are dubious.

7.4 The Misuse of Anonymity Protections with Template Plagiarism

Qualitative research on sensitive or controversial issues often necessitates that confidentiality and anonymity protections be granted to interviewees. These protections are important, since interviewees are more likely to be candid in their responses if they are confident that their identities will be protected. Researchers must insure that their published findings cannot be traced back to their interviewees. According to one classic formulation regarding anonymity protections, "data should be presented in such a way that respondents should be able to recognise themselves, while the reader should not be able to identify them" (Grinyer 2002: 1). One downside to these protections is that qualitative data sets are typically unverifiable by third parties, who are left to trust that veracity of what is being reported. Proving that a researcher's publications are the product of fabricated, fraudulent, or inaccurate fieldwork data is typically not possible. In one rare case involving a European political anthropologist, however, a 2013 investigating committee found claims of fabricated fieldwork data to be "highly plausible" when no evidence of the sources could be discovered (Baud et al. 2013: 17). The researcher was found to have committed scientific misconduct, and some of publications later received retractions from editors and publishers (de Boer and Keulemans 2013; Ferguson 2014). A more recent of discovery of influential fieldwork that was revealed to be unreliable comes the field of psychology (Calahan 2019).

It is not easy to discredit qualitative data reported to have been collected under the conditions of confidentiality or anonymity. Nevertheless, when what is represented as an anonymized data set is really the product of template plagiarism, a comparison of a suspected source text with the suspected plagiarizing text can disclose sufficient idiosyncratic similarities to prove scientific misconduct. There are some additional book chapters and journal articles by C. that appear to be the product of template plagiarism, with the additional element that they contain lengthier methodological discussions of how the putative fieldwork data set was obtained.

7.4.1 Triplicate Publication and Template Plagiarism

To take an extremely usual case: the Brill quarterly *International Journal on Minority and Group Rights* published three versions of the same article by C. in three consecutive years (C. 2017a, 2018, 2019b). The penultimate and ultimate versions contained additional quotation marks and references, and the 2019 version was accompanied with an erratum explaining, "a number of footnotes and quotation marks were missing" (C. 2019b: 5). In its three iterations, the author of record explains the purported research methodology, stating "we complement our study by adding semi-structured interviews with three representatives of NGOs working in the social area in Belgrade as a qualitative research method" (C. 2017a: 124; see C. 2018: 459, 2019b: 307).

Table 7.13 Three versions of a passage alongside its suspected source text

C. 2017a: 141	C. 2018: 476	C. 2019b: 326	Pejčić 2007: 61–62
According to a local representative of an NGO in Belgrade, the Roma community in Serbia, as a minority group under the burden of historical discrimination: "<u>are</u> still <u>facing institutional, structural and individual discrimination. Roman</u>ians <u>have limited access to educational and employment possibilities, lack of appropriate housing, insufficient healthcare as well as limited access to social services and public life.</u>"[50]	According to a local representative of an NGO in Belgrade, the Roma community in Serbia, as a minority group under the burden of historical discrimination, "<u>are</u> still <u>facing institutional, structural and individual discrimination. Roman</u>ians <u>have limited access to educational and employment possibilities, lack of appropriate housing, insufficient healthcare as well as limited access to social services and public life</u>".[49]	According to a local representative of an NGO in Belgrade, the Roma community in Serbia, as a minority group under the burden of historical discrimination, "<u>are</u> still <u>facing institutional, structural and individual discrimination</u>".[80] [80] Interview with a local representative of an NGO in Belgrade in April 2015.	The Roma community in Serbia, as a minority group under the burden of historical discrimination, <u>are facing institutional, structural and individual discrimination. Roman</u>ies <u>have limited access to educational and employment possibilities, lack of appropriate housing, insufficient healthcare as well as limited access to social services and public life.</u>
[50] Interview with a local representative of a NGO in Belgrade in April 2015.	[49] Interview with a local representative of an NGO in Belgrade in April 2015.		

The article specifies further in several footnotes that the interviews were carried out "in Belgrade in April 2015."

The cloak of anonymity bestowed upon the purported interviewees results in very little being known about them. They are only identified collectively as three NGO representatives and are not distinguished individually when quotations are attributed to them. Five quotations are ascribed to the three interviewees in the course of the 2017 and 2018 versions, but how the five quotations divide among the three is unspecified. In the 2019 version, three of the quotations are omitted entirely from the paper, and of the two that remain, one is considerably shortened. Table 7.13 displays one quotation common to all three versions, alongside the source text for the suspected template plagiarism.[2]

By publishing the article three times in consecutive volumes of *International Journal on Minority and Group Rights* over the period 2017–2019, C. has warranted the veracity of this particular fieldwork quotation no less than three times. Nevertheless, it appears that the quotation has been produced using a source text as a template. The apparent source text is a 2007 master's thesis by Mirella Pejčić from Uppsala University. There is no doubt that C. is familiar with this thesis, as C. cites it elsewhere in all three versions of the article. But here a quotation consisting of 34 words in the 2017 and 2018 versions of the article ascribed to the anonymized NGO representative is found verbatim in Pejčić's master's thesis, with the slight alteration that the word "still" has been added and the word "Romanies" has been modified as "Romanians." This 34-word quotation is inexplicably trimmed in the 2019 version, but still the substantial overlap with the suspected source text is evident.

To be clear: *the words C. represents as having been uttered by an NGO representative in Belgrade in April 2015 overlap with a master's thesis that was issued*

---

[2]For a complete analysis of the fieldwork data in the article published three times in *International Journal on Minority and Group Rights* as C. 2017a, 2018, 2019b, see Dougherty (2020).

7.4 The Misuse of Anonymity Protections with Template Plagiarism 145

8 years before the interview was alleged to have taken taken place. In the thesis, the words appear as part of Pejčićs original analysis, not as data obtained from an interview. Crucially significant in support of a finding of template plagiarism is the fact that the sentence introducing the putative quotation in C. 2017a, 2018, 2019b is also found in Pejčić's master's thesis, on the same page ("The Roma community in Serbia, as a minority group under the burden of historical discrimination […]").

One must ask how the identity of the passage in the three versions by C. and the master's thesis by Pejčić could be possible, since the identity is not only extensive but it is also idiosyncratic. The oddity should not be understated. If the professed qualitative data set for C. 2017a, 2018, 2019b is to be trusted, then one must take as possible a scenario where an anonymized interviewee working for an NGO in Belgrade in 2015 is expressing herself or himself in verbatim propositions that overlap with sentences published eight years earlier in a relatively obscure master's thesis from a Swedish university. What is more, it must further be believed that in introducing the qualitative data, C. is also formulating utterances that correspond verbatim with the same thesis. It seems more plausible to assume that the passage published three times in C. 2017a, 2018, 2019b is an instance of template plagiarism based in part on the 2008 Swedish master's thesis. The data set seems unreliable. Nevertheless, there are positive recent citations in the downstream literature to the 2017 and 2018 versions of the article.

7.5 Sources for Template Plagiarism

The examples of suspected data fabrication by means of template plagiarism discussed above have involved a range of undisclosed source texts. They have included:

- EU reports (Tables 7.7, 7.8)
- a newspaper article (Table 7.9)
- a journal article (Table 7.10)
- a conference presentation (Table 7.11)
- a newspaper editorial (Table 7.12)
- an unpublished master's thesis (Table 7.13).

Passages from these diverse texts appear to have been transformed into (1) on-the-ground field interview quotations as well as (2) the surrounding text that introduces them. In some disciplines, fieldwork interviewing is regarded as a valuable research method because interviewees provide insights into topics under investigation that could not otherwise be acquired. The consideration of the unique perspectives offered by on-the-ground interviewees can advance what is known in some areas of research. In the cases discussed above, however, readers are lead to believe they are being presented such unique perspectives from interviewees, but it appears that the perspectives are not the product of fieldwork interviews but instead the considered view of outsiders commenting on states of affairs in the Western Balkans from a variety of perspectives. The repackaging of outsider views in the guise of first-person accounts denies readers of C.'s articles the perspective they are told they are being offered.

More troubling, however, is the apparent denial of a voice to vulnerable populations covered in the articles. If the reported fieldwork quotations are unreliable, then true spokespersons of vulnerable populations are not represented in articles that are meant to form the basis of policy recommendations for these populations. One benefit of properly performed qualitative research is to give a voice to marginalized populations whose perspectives would otherwise be missed through other forms of research.

7.6 A Question of Scale

The examples of suspected template plagiarism discussed in this chapter were selected from the body of published work from one author of record—C.—in publications that appeared in print from 2007–2019. They included examples of suspected standard template plagiarism, template text recycling, and template plagiarism involving suspected data fraud. The affected journals and publishing houses represent a major sector of the scholarly publishing industry. Publishers include Elsevier, Routledge, Brill, Taylor & Francis, and Wolters Kluwer, among others. If one potential malefactor is able to land plagiarizing work in so many respected venues, and some quite recently, one can draw the conclusion that the inability to guard against disguised plagiarism must be prevalent.

The examples here are meant to be illustrative of template plagiarism, and should not be taken to mean that the problem is restricted to one field. The phenomenon is present in publications from many disciplines and is practiced by authors of record who publish in a variety of venues. To give one example: in 2015 a researcher ("F") working in the area of economic history began receiving retractions for plagiarism (Palus 2015). Table 7.14 shows one example illustrating how template plagiarism was used to produce the illusion of new research.

The source text (Riley 1982) is an analysis of social contract theory found in major early modern philosophers—Hobbes, Locke, Rousseau, and Kant—and has been taken as a template to produce an analysis of Spanish economic thought from the sixteenth and seventeenth centuries. The retraction statement for F. 1998 explained that "direct reference and citation of the works of other scholars is often inconsistent and in some cases totally lacking" (Brill 2015).

Table 7.14 Template plagiarism in economic history

F. 1998: 534	Riley 1982: 22
Contract theory appears to require a notion of will as a moral causality that produces legitimacy, obligation, and the like, as its effects […]. To what extent does the scholastic analysis and validation of economic exchange depend on their understanding of the contract's theory? Do the Spanish scholastic doctors provide a coherent account of the moral causality?	contract theory appears to require a notion of will as a moral causality that produces legitimacy, obligation, and the like, as effects. The question is: to what extent do Hobbes, Locke, Rousseau, and Kant provide a coherent account of that "causality"?

Humanities disciplines may not be the ones most affected by template plagiarism. Recent work has focused on the use of templates in the production of fraudulent articles in biomedical fields (Byrne and Christopher 2020). In these fields, the use of templates is often accompanied by data fabrication. Some of these defective articles have been are produced by paper mills that were offered for sale to researchers

7.7 Conclusion

In 2019, I sent retraction requests sent to editors and publishers for 13 articles for which C. is the author (or co-author) of record, including those discussed in this chapter (C. 2009, 2011, 2013a, b, c, 2014a, 2014b/2019, 2016, 2017a, b, 2018, 2019c). The requests were made on the basis of suspected (1) copy-and-paste plagiarism, (2) template plagiarism, and/or (3) data fabrication. At the time, four other articles by C. had already been retracted for plagiarism and a failure to credit sources properly (see Oransky 2019).

European Foreign Affairs Review retracted C. 2013b in September 2019, stating "the publisher and the editors [...] have received proof of a serious lack of academic rigor" and that "in response to this finding, a clear breach of our policies on professional ethics and against our high integrity standards, it has been decided to retract the relevant article" (European Foreign Affairs Review 2019). Likewise, in November 2019, *Journal of Contemporary European Studies* issued a retraction for C. 2013c that stated the article "has been retracted and should not be cited" (Journal of Contemporary European Studies 2020: 138). The retraction explained that "it has been determined that the paper contains significant passages of unattributed material, as well as material which paraphrases source material with only minor alterations" (ibid.). The journal could not confirm whether the purported fieldwork interviews ever took place, and the editor and publisher retracted the article on the basis of the plagiarism (see Ascensão 2019; Filipa 2019; País ao Minuto 2019; Jornal de Notícias 2019).

Retractions for C. 2011 and C. et al. 2016 were published in *RBPI* in early 2020 and each stated that "This article is being retracted due to the identification of passages without due citation and others reproduced verbatim without quotation marks nor attribution" (RBPI 2020a, b). The publisher Brill issued a retraction for C. 2016/2018 also in early 2020, explaining that support had been found for a suspicion of plagiarism and "The editors deemed the article unacceptable for publication" (Brill 2020). A retraction for C. 2014b/2019 is forthcoming from Brill. Lexington Books, the publisher of the volume edited by C. and that contained C. 2014a, placed the volume on hold in August 2019 after receiving evidence of suspected template plagiarism, and then removed the book from print two months later. Routledge removed the book containing 2013a from sale in both the print and electronic formats in November 2019, seven months after received a request for retraction.

References

Abylaev, Mansur, et al. 2014. Resilience challenges for textile enterprises in a transitional economy and regional trade perspective. *International Journal of Supply Chain and Operations Resilience* 1 (1): 54–75. https://doi.org/10.1504/IJSCOR.2014.065459.

Ağir, Bülent Sarper. 2009. Rethinking Security in the Balkans. *SDU Faculty of Arts and Sciences Journal of Social Sciences*. https://dergipark.org.tr/download/article-file/117956.

Anastasijevic, Dejan. 2006. *Organized crime in the Western Balkans*. https://www.files.ethz.ch/isn/102340/1_Anastasijevic.pdf.

Ascensão, Joana. 2019. Professora universitária acusada de plágio. *Observador*, November 26. https://observador.pt/2019/11/26/professora-universitaria-acusada-de-plagio-pelo-menos-5-artigos-retirados.

Baud, Michiel, et al. 2013. *Circumventing reality*. Amsterdam. www.vu.nl/en/Images/20131112_Rapport_Commissie_Baud_Engelse_versie_definitief_tcm270-365093.pdf.

de Boer, Richard, and Maarten Keulemans. 2013. Oud-hoogleraar Bax schuldig aan wetenschappelijk 'wangedrag'. *De Volkskrant*, September 23. www.volkskrant.nl/nieuws-achtergrond/oud-hoogleraar-bax-schuldig-aan-wetenschappelijk-wangedrag~b6c16396.

Bouville, Mathieu. 2008. Plagiarism: Words and ideas. *Science and Engineering Ethics* 14 (3): 311–322.

Brill. 2015. *It has been brought to our attention [...]*. Leiden: Brill. https://brill.com/view/title/159.

Brill. 2020. Retraction notice. *Southeastern Europe*. https://doi.org/10.1163/18763332-0401025A.

Byrne, Jennifer A., and Jana Christopher. 2020. How can journals and peer reviewers detect manuscripts and publications from paper mills? *FEBS Letters* 594 (4): 583–589.

Bull, Martin J., and James L. Newell. 2003. Conclusion: Political corruption in contemporary democracies. In *Corruption in contemporary politics*, ed. Martin J. Bull and James L. Newell, 234–247. Basingstoke: Palgrave Macmillan.

Chaddah, Praveen. 2014. Not all plagiarism requires a retraction. *Nature* 511 (7508): 127. https://doi.org/10.1038/511127a.

Calahan, Susannah. 2019. *The Great Pretender*. New York: Hachette.

[C.], et al. 2007. Globalization, regionalization, and Europeanization. In *Globalization: Perspectives from Central and Eastern Europe*, ed. Katalin Fábián, 173–195. Amsterdam: Elsevier.

[C.]. 2009. Europeanization impact on Croatia's course to democracy. *Nação e Defesa* 122: 173–201. https://www.idn.gov.pt/index.php?mod=1321&cod=125#sthash.n4WDknMu.dpbs.

[C.]. 2011. Human rights promotion in Serbia. *RBPI* 54 (1): 142–158. [Retracted in 2020 in 63 (1)].

[C.]. 2013a. Threats to human security in the Western Balkans. In *The European Union neighbourhood*, ed. [C.], 33–53. Farnham: Ashgate. [Republished in 2016. New York: Routledge] [Removed from sale due to plagiarism].

[C.]. 2013b. European Union security policy in the former Yugoslav Republic of Macedonia. *European Foreign Affairs Review* 18 (3): 429–447. [Retracted in: European Foreign Affairs Review 2019.].

[C.]. 2013c. Civil society in Macedonia's democratization process. *Journal of Contemporary European Studies* 21 (2): 202–217. [Retracted in 2019].

[C.]. 2014a. Croatia's difficult path to European Union membership. In *European institutions, democratization, and human rights protection in the European periphery*, ed. Henry F. Carey 245–270. Lanham: Lexington Books. [Removed from sale due to plagiarism].

[C.] 2014b/2019. Albania's difficult path towards democracy. *Canadian-American Slavic Studies* 48 (4): 468–491. [Retraction forthcoming].

[C.]. 2016. Bridging the gap. *Journal of Contemporary Central and Eastern Europe* 24 (2): 113–129.

[C.]. 2016/2018. The security sector reform in Macedonia. *Southeastern Europe*: 1–25. [Retracted in 2020].

[C.] et al. 2016. The European Union and the member states. *RBPI* 59 (1): 1–18. [Retracted in 2020 in 63 (1)].

[C.]. 2017a. The limits of Europeanization on minority rights in Serbia. *International Journal on Minority and Group Rights* 24 (2): 123–149.

[C.]. 2017b. Promoting the rule of law in Serbia. *Communist and Post-Communist Studies* 50 (4): 331–337.

[C.]. 2018. The limits of Europeanization on the minority rights in Serbia. *International Journal of Minority and Group Rights* 25 (3): 458–484.

[C.]. 2019a. Correction. *Journal of Contemporary European Studies* 27 (1): iii–cii.

[C.]. 2019b. Erratum to The Limits of Europeanization on the Minority Rights in Serbia: The Roma Minority [25:3 International Journal of Minority and Group Rights (2018) pp. 458–484], *International Journal on Minority and Group Rights* 26 (2): 305–334.

[C.]. 2019c. The European Union accession and climate change policies in the Western Balkan countries. In *Climate change and global development*, ed. Tiago Sequeira and Liliana Reis, 153–173. Cham: Springer.

Dallara. Cristina. 2014. *Democracy and judicial reforms in south-east Europe*. Cham: Springer.

Dougherty, M.V. 2018. *Correcting the scholarly record for research integrity*. Cham: Springer.

Dougherty, M. V. 2020. A concern about qualitative research in two recent articles. *International Journal on Minority and Group Rights* 27, Forthcoming.

European Foreign Affairs Review. 2019. To whom it may concern. Wolters Kluwer. https://www.kluwerlawonline.com/abstract.php?area=Journals&id=EERR2013026.

[F.] 1998. Later Scholastics: Spanish economic thought in the XVIth and XVIIth centuries. In *Ancient and medieval economic ideas and concepts of social justice*, ed. S. Todd Lowry and Barry Gordon, 503–561. Leiden: Brill, 1998. [Retracted in: Brill 2015].

Fosztó, László. 2018. Recent publications. *Newsletter of the G. L. Society* 41 (2): 8–15.

Grinyer, Anne. 2002. The anonymity of research participants. *Social Research Update* 36: 1–4. http://sru.soc.surrey.ac.uk/SRU36.PDF.

Goldingay, Sarah. 2009. Plagiarising theory. *Studies in Theatre and Performance* 29 (1): 5–14.

Horbach, S.P.J.M., and W. Halffman. 2019. The extent and causes of academic text recycling or 'self-plagiarism'. *Research Policy* 48 (2): 492–502.

Hope, Sr, Kempe Ronald, and Bornwell C. Chikulo (eds.). 2000. *Corruption and development in Africa*. New York: St. Martin's.

[J.]. 2013. The Europeanization impact on Albania course to democracy. In *Democracy in times of turmoil*. https://www.sociology.al/sites/default/files/8th_International_Conference_2013_Proceedings.pdf.

Jayasuriya, Nalin. 2012. Occupy everything. *Daily FT*, January 11. http://www.ft.lk/columns/occupy-everything/4-65344.

Jornal de Notícias 2019. Professor norte-americano alerta para plágio de professora na Universidade do Porto. *Jornal de Notícias*, November 27. https://www.jn.pt/nacional/universidade-do-porto-pode-levar-caso-da-professora-acusada-de-plagio-a-comissao-de-etica-11559944.html.

Journal of Contemporary European Studies. 2020. Statement of retraction. *Journal of Contemporary European Studies* 28 (1): 138.

Lavdas, Kostas. A. 2017. From authoritarianism to Europeanization. *European Quarterly of Political Attitudes and Mentalities*, 6 (2): 101–115. https://www.ssoar.info/ssoar/handle/document/51770.

Mandazhieva, Petya. 2009. *European way of doing security*. Master's thesis. Central European University, Budapest. http://www.etd.ceu.hu/2009/mandazhieva_petya.pdf.

Marušić, Tea. 2018. A new publishing approach—retract and replace—is having growing pains. *Retraction Watch*, March 29. https://retractionwatch.com/2018/03/29/a-new-publishing-approach-retract-and-replace-is-having-growing-pains/.

Marasović, Tea, et al. 2018. Transparency of retracting and replacing articles. *The Lancet* 391 (10127): 1244–1245.

Mitsilegas, Valsamis, et al. 2003. *The European Union and internal security*. Basingstoke: Palgrave Macmillan.

Oransky, Ivan. 2019. Political science prof up to five retractions. *Retraction Watch*, November 25. https://retractionwatch.com/2019/11/25/political-science-prof-up-to-five-retractions-for-plagiarism.

País ao Minuto. 2019. Universidade pode levar caso de professora acusada de plágio a comissão. *Notícias ao Minuto*, November 27. https://www.noticiasaominuto.com/pais/1366828/universidade-pode-levar-caso-de-professora-acusada-de-plagio-a-comissao.

Palus, Shannon. 2015. Two retractions cost economic historian book chapter and journal article. *Retraction Watch*, July 20. http://retractionwatch.com/2015/07/20/two-retractions-cost-economic-historian-book-chapter-and-journal-article.

Pejčić, Mirella. 2007. *Minority rights in Serbia*. Master's Thesis, Uppsala University. https://www.pcr.uu.se/digitalAssets/654/c_654492-l_1-k_mfs_pejcic.pdf.

Pemberton, Michael, et al. 2019. Text recycling. *Learned Publishing* 32 (4): 355–366.

RBPI. 2020a. Retraction. *RBPI* 63 (1). https://doi.org/10.1590/0034-73292011000100009retract.

RBPI. 2020b. Retraction. *RBPI* 63 (1). https://doi.org/10.1590/0034-7329201600103retract.

Riley, Patrick. 1982. *Will and political legitimacy*. Cambridge: Harvard University Press.

RTV B92. 2013. Serbia 72nd on corruption perception list, December 4. https://www.b92.net/eng/news/business.php?yyyy=2013&mm=12&dd=04&nav_id=88546.

[J.] The Europeanization impact on Albania course to democracy. In *Democracy in times of turmoil*. https://www.sociology.al/sites/default/files/8th_International_Conference_2013_Proceedings.pdf.

Reinhard, Janine. 2008. EU democracy promotion through conditionality. http://kops.uni-konstanz.de/handle/123456789/3959.

Roig, Miguel. 2015. *Avoiding plagiarism, self-plagiarism, and other questionable writing practices*, 2nd ed. USDHHS, ORI. https://ori.hhs.gov/sites/default/files/plagiarism.pdf.

Schechner, Richard, et al. 2009. Concerning theory for performance studies. *The Drama Review* 53 (1): 7–46.

Shani, Giorgio. 2007. Introduction: Protecting human security in a post 9/11 World. In *Protecting human security in a post 9/11 World*, ed. Giorgio Shani et al., 1–14. Basingstoke: Palgrave Macmillan.

Silva, Filipa. 2019. Professora da FLUP sob suspeita de plágio. *JPN-JornalismoPortoNet*, November 27. https://jpn.up.pt/2019/11/27/professora-da-flup-sob-suspeita-de-plagio.

St. Onge, K.R. 1988. *The melancholy anatomy of plagiarism*. Lanham: University Press of America.

Stojanova, Ana. 2013. *Elite- or mass-driven democratic consolidation?* Ph.D. dissertation. IMT Institute for Advanced Studies, Lucca. http://e-theses.imtlucca.it/125/1/Stojanova_phdthesis.pdf.

Tafili, Najada. 2008. Consolidation of democracy: Albania. *Journal of Political Inquiry* 1: 1–13.

TACSO. 2011. *Needs assessment Former Yugoslav Republic of Macedonia*. October 2011. Skopje http://tacso.eu/wp-content/uploads/2019/05/mk_nar_sep2011.pdf.

Uğurlu, Sezen. 2013. Political criteria for accession to the EU. *SBF Dergisi* 68 (3): 165–190. https://dspace.ankara.edu.tr/xmlui/bitstream/handle/123456789/53270/19144.pdf?sequence=1.

Weber-Wulff, Debora. 2014. *False feathers*. Heidelberg: Springer.

Chapter 8
Conclusion: Remedies for Disguised Plagiarism

Abstract This short chapter concludes a book-length study of under-recognized forms of disguised plagiarism. All varieties of disguised plagiarism possess some additional concealment beyond the mere copying of words with inadequate attribution. The book focused on six types of disguised plagiarism: translation, compression, dispersal, magisterial, exposition, and template plagiarism. This concluding chapter identifies resources that assist editors, publishers, and whistleblowers in maintaining the integrity of the body of published research.

Keywords Disguised plagiarism · Retractions · Whistleblowing · Post-publication peer review

In all varieties of disguised plagiarism, additional concealment creates a greater distance between the source text and the copy. The previous chapters defended a typology of six principal forms: translation, compression, dispersal, magisterial, exposition, and template plagiarism. As the case studies from the scholarly research literature in the humanities reveal, instances of plagiarism can be complex, especially when an article combines disguised forms of plagiarism. When disguised forms are blended, the identification of plagiarizing articles is especially difficult. The plagiarism in articles that combine compression plagiarism and translation plagiarism, or combine template and exposition plagiarism, may remain undiscovered for years or even in perpetuity. The likelihood is high that the readership of such articles will unwittingly regard them as trustworthy.

Even if one were to suspect that a given article is the product of one or more forms of disguised plagiarism, providing proof of an article's inauthenticity is challenging. The identification of various kinds of evidence, and the ability to display the evidence in clear, persuasive, and verifiable ways, are important skills that should be possessed by whistleblowers, editors, and publishers. Strong case building is typically a prerequisite for securing a correction of the published research literature. The many examples provided in this book illustrate what counts as evidence and exhibit diverse ways for demonstrating disguised plagiarism.

Plagiarism in all of its forms—disguised and undisguised—corrupts the body of published research literature, making it less reliable for students and researchers who depend upon it. Acts of disguised plagiarism are especially pernicious; they are much more difficult to extirpate. Undetected plagiarism denies credit to original authors for their research findings and falsifies the history of discovery. Acts of disguised plagiarism are potentially more disruptive of the larger academic enterprise than instances of garden-variety copy-and-paste plagiarism. Since acts of disguised plagiarism are likely to persist undetected for longer periods of time, they distort a plagiarist's publication profile with greater influence. Due to undetected plagiarism, grants, promotions, job offers, awards, and academic honors are bestowed on the undeserving and thus denied to genuine researchers. Academic plagiarism not only damages the reliability of the body of published research, but it creates inefficiencies for researchers and the institutions that support research.

Corrections of the published research literature take many forms, but the principal ones (in order of gravity) are: retractions, errata, corrigenda, and expressions of concern (Hames 2007: 193–195; Boxheimer and Pulvere 2019: 1–3). A retraction is the optimal correction of a plagiarizing article; it is an official notification issued by a publisher that declares that the article is not reliable. The most successful retractions for plagiarism are promulgated widely to readers and give credit to those authors whose works have been misappropriated. Less successful retractions neglect to identify the source texts, are hidden behind paywalls, and are not linked electronically to the retracted articles. Some publishers reserve retractions for only major cases of plagiarism, judging that that minor instances of plagiarism in small portions of an article can be corrected through the issuance of a less grave correction (e.g., corrigendum) rather than a retraction (see Wager 2013). The precise form that published corrections should take for varieties of research misconduct—not just plagiarism—continues to be an area of continued debate (see Barbour et al. 2017; Fanelli et al. 2018). Products of disguised plagiarism are among the strongest candidates for retraction.

8.1 Resources and Tools

Despite the damage that plagiarism in all of its varieties inflicts on the body of published research literature, there are reasons to be optimistic about solving the problem of plagiarism. The appearance of lengthy studies on plagiarism indicates that the researchers and publishers are becoming more vigilant. Specialized books focusing on fields in the sciences (Zhang 2016) as well as in the humanities (Dougherty 2018) complement earlier groundbreaking work that considered plagiarism in European dissertations (Weber-Wulff 2014). Text-matching software is steadily improving, and free online text-comparison tools assist investigators in confirming some forms of textual overlap (Quelle: Textvergleich, n.d.; Similarity Texter, n.d.). The journalists at Retraction Watch have created a database for retractions featuring over 20,000 entries, allowing researchers to check that status of articles (Retraction Watch Database 2015). This resource addresses the problem that retracted articles continue to acquire positive citations in the downstream literature.

The influence of post-publication peer review has expanded through PubPeer, an online platform that allows readers to raise concerns about the integrity of published research articles. The online group Vroniplag Wiki crowdsources examinations of texts for plagiarism, focusing principally on university theses and dissertations. The Committee on Publication Ethics continues to produce flowcharts that advise editors and publishers for how to deal with various research integrity violations, including plagiarism.

Violations of the accepted principles of research should be discussed openly, and evidence of suspected research misconduct should be examined thoroughly. The English words publish and publication are cognates of the Latin *publicare*, "to make public" (Glare 1982: 1512). When inauthentic researchers make public their plagiarizing work by publishing it, a public examination is warranted. Post-publication peer review of plagiarizing articles not only generates support for the needed retractions, but also allows members of the of the research community to become acquainted with the various subtle forms that research misconduct can take. The careful scrutiny of works for evidence of suspected misconduct is a necessary task in the enterprise of research.

References

Barbour, Virginia, et al. 2017. Amending published articles. *F1000Research* 6:1960. https://doi.org/10.12688/f1000research.13060.1.
Boxheimer, Erica Wilfong and Bernd Pulvere. 2019. Self-correction prevents withdrawal syndrome. *The Embo Journal* 38: e70001. https://doi.org/10.15252/embj.201970001.
Dougherty, M.V. 2018. *Correcting the scholarly record for research integrity*. Cham: Springer.
Fanelli, Daniele, et al. 2018. Improving the integrity of published science. *European Journal of Clinical Investigation* 48 (4): 1–6, e12898.
Glare, P.G.W. (ed.). 1982. *Oxford Latin dictionary*. Oxford: Clarendon Press.
Hames, Irene. 2007. *Peer review and manuscript management in scientific journals*. Malden: Blackwell Publishing.
Quelle: Textvergleich. n.d. https://vroniplag.wikia.org/de/wiki/Quelle:Textvergleich.
Retraction Watch Database. 2015. http://retractiondatabase.org.
Similarity Texter. n.d. https://people.f4.htw-berlin.de/~weberwu/simtexter/app.html.
Wager, Liz. 2013. Suspected plagiarism in a published manuscript. COPE. https://publicationethics.org/resources/flowcharts/suspected-plagiarism-published-manuscript.
Weber-Wulff, Debora. 2014. *False feathers*. Heidelberg: Springer.
Zhang, Yuehong (Helen). 2016. *Against plagiarism*. Cham: Springer.

Index

A
Africa, 129, 130
Albania, 129, 134, 135
Allen, John, 94, 95
Anderson, Mark, 123
Anderson, Mary, 90, 91, 95
Antiphon, 77, 83, 84
Anti-plagiarism software, *see* Text-matching software
Apologies, 47
Argument, 2, 7–9, 16, 19–22, 37, 39, 46, 58, 59
Argumentation, 5, 37, 38, 40–47, 61
Attribution, deficient, 105, 108, 110, 111, 116, 123
Authority
   Magisterial, 5, 75, 84, 91
   of canonical authors, 3
   of exegetes, 6, 103, 125
Autobiography, *see* meta-narrative

B
Benedict XVI, 80, 87, 88, 90
Berichte zur Wissenschaftsgeschichte, 104
Biagioli, Mario, 59
Bibliography, 2, 55, 68, 105, 117, 118, 123, 124
Bibliometrics, 51, 67
Blumenbach, Johann Friedrich, 110
Brill, 47, 143, 146
British Journal for the History of Science, 123

C
Cambridge Texts in the History of Philosophy, 116
Cambridge University Press, 54, 56
Canonical texts, 6, 103, 116, 117, 120, 121, 125
Cardinal Marc Ouellet, 81, 82, 84, 85, 87, 89, 90, 92, 93, 95, 96, 98
Cardinal William Levada, 76, 77, 79, 81, 83, 86, 92, 96
Carter, Harry, 18, 19
Committee on Publication Ethics, 153
Communication, scholarly
   corruption of, 16, 24, 43
Compression Plagiarism, *see* Plagiarism, Compression
Concealment, 2, 5, 8, 17, 20, 23, 59, 69, 79, 82, 92, 93, 99, 110, 139
Copyright, 56
Corrigenda, 3, 44, 68, 84, 152
Croatia, 129, 131, 134, 135

D
Danesi, Marcelo, 24
Data
   fabrication of, 6, 135, 137, 145, 147
   qualitative, 136, 138, 139, 142, 143, 145
Disguised plagiarism, *see* Plagiarism, Disguised
Dougherty, M. V., 9, 30, 34, 44, 47, 62, 63, 66, 67, 92, 134, 136, 144, 152
Driscoll, Jeremy, 77–86, 96, 97
Duchesneau, François, 104
Dupont, Jean-Claude, 108, 109, 124

E
Errata, 3, 44, 123, 143, 152
European Foreign Affairs Review, 133, 147
European Union (EU), 128, 133, 136–138, 145
Exposition plagiarism, *see* Plagiarism, Exposition
Expressions of Concern (EoC), 53, 64, 152

F
Fabrication, 6, 14, 127, 135, 137, 145, 147
Fahnestock, Jeanne, 43
Feix, Josef, 18–23
First-person narrative, *see* meta-narrative
Francis, Pope, 93, 99
Friedman, Russell, 62, 63, 66

G
Ghostwriting, 76, 81, 93, 98, 99
Ginsborg, Hanna, 112, 124
Gipp, Bela, 2, 13–15, 41, 94
Goati, Vladimir, 139
Godley, A. D., 18, 19, 21–23, 26
Gosepath, Stefan, 25–32, 34, 38–44, 46, 47

H
Haller, Albrecht von, 110, 111, 121, 124, 125
Harsting, Pernille, 18, 62, 63, 66, 68
Hegel, G. W. F., 104, 117, 124, 125
Henry of Ghent, 64
Herodotus, 16, 18–25
Heythrop Journal, The, 54–56, 58
Historiography, 103–105, 123
Historisches Wörterbuch der Philosophie, 26
History, 1, 16, 62, 84, 104, 111, 116, 119, 123, 125, 128, 146, 152
History and Philosophy of the Life Sciences, 123
Hochschild, Joshua, 92
Humanities, 1–3, 13, 34, 147, 151, 152
Hume, David, 53, 57
Hunfield, Hans, 21
Hutton, Christopher, 61

I
Idea plagiarism, *see* Plagiarism, Idea
International Association for Semiotic Studies, 69
International Journal of Law in Context, 54

International Journal on Minority and Group Rights, 143, 144
International Society for the History of Philosophy of Science, 104

J
John Paul II, 86, 90, 95
Journal of Contemporary Central and Eastern Europe, 139
Journal of Contemporary European Studies, 136, 138, 147

K
Kant, Immanuel, 22, 104, 108, 112, 116, 117, 125, 146
Kantola, Ilkka, 62–65, 71
Katholischen Universität Eichstätt-Ingolstadt, 21
Kielmeyer, Carl Friedrich, 110, 117
Kribbe, Hans, 53–62, 64, 71

L
Leibniz, Gottfried Wilhelm, 118
Lenoir, Timothy, 112–114, 125
Libraire Droz, 32
Literal plagiarism, *see* Plagiarism, Literal
Locke, John, 53, 57, 146
London School of Economics and Political Science, 53
Luther-Agricola-Society, 62

M
Macedonia, 129, 131, 136, 137
Mad Libs, 127
Magisterial Plagiarism, *see* Plagiarism, Magisterial
Magisterial texts
  apostolic exhortation, 75, 93
  encyclical, 75
  general address, 75
  homily, 95
  pastoral letter, 75
Magisterium, 75, 80
Martin, Ben, 14
Meta-narrative, 26, 27, 31, 38, 85, 105
Miller, A. V., 117, 124
Mind, 32
Misconduct, research, *see* Research misconduct
More, Thomas, 22

Index 157

N
Nação e Defesa, 134
Netherlands Organisation for Scientific Research, 38
Notre Dame Philosophical Reviews, 104

O
Ohio Dominican University, 129
Oken, Lorenz, 118–121, 124, 125
Optical Character Recognition (OCR) errors, 58
Origins, 5, 7, 8, 14, 23, 59, 61, 76, 79–81, 83, 84, 87, 92, 105
Orthography
    American English, 131, 140
    British English, 121, 131, 140
    German, 19
Oxford University Press, 33

P
Parfit, Derek, 53
Paul VI, 75
Pawn-sacrifice plagiarism, *see* Plagiarism, Pawn-sacrifice
Peacocke, Christopher, 33
Peterson, Keith, 114, 115, 124
Philosophy, plagiarism in, 1, 13, 16, 32, 38, 53, 70, 71
Plagiarism
    compression, 3–5, 37, 38, 40–44, 46, 47, 52, 82, 117, 151
    copy-and-paste, 2, 9, 38, 44, 131, 137, 142, 147, 152
    disguised, 1–5, 9, 10, 13, 14, 18, 33, 37, 38, 46, 52, 75, 125, 127, 128, 130, 137, 146, 151, 152
    double, 81, 83
    exposition, 3, 6, 103, 117, 122, 151
    idea, 2
    llteral, 2
    magisterial, 3, 5, 75, 89, 92, 99
    patchwork, 86, 98
    pawn sacrifice, 2, 109
    self-, 2, 3, 8, 134
    serial, 4, 10, 44, 62, 66, 92, 94, 97
    shake-and-paste, 2
    structural, 2
    template, 3, 6, 78, 127, 128, 130–137, 139, 141–147, 151
    translation, 3, 4, 13–16, 18, 22, 24, 25, 29, 32–34, 41, 44, 52, 67, 151
    triple, 96
    typology of, 2, 3, 10, 41, 44
    verbatim, 94
Plagiarists, 1–10, 37, 38, 41, 46, 51–53, 59, 67, 81, 94, 95, 97, 98, 127, 152
Plato, 16, 67, 69–71
Poland, 134, 135
Political science, plagiarism in, 1, 127, 128
Post-Publication Peer Review (PPPR), 32, 38, 56, 104, 153
Pro Ecclesia, 77, 83, 84
PubPeer, 32, 33, 56, 68, 153

R
Ratio Juris, 54–56
Ratzinger, Joseph, *see* Benedict XVI
Readership
    deceived, 7
Reinhold Treviranus, Gottfried, 117
Relay translation, 22, 23
Research literature, *see* Scholarly Record
Research misconduct, 2–4, 6, 14, 15, 54, 67, 71, 127, 132, 152, 153
Responding articles, 7–9, 24, 32, 43, 47
Retractions, 1, 3, 4, 6, 7, 9, 14, 15, 32, 34, 38, 44, 47, 52, 53, 55, 56, 61, 62, 64, 66, 67, 92, 93, 123, 134, 143, 146, 147, 152, 153
Revista Brasileira de Política Internacional, 133
Rigotti, Eddo, 15, 24, 25
Rocci, Andrea, 24
Roe, Shirley, 107–109, 124
Routledge, 128, 141, 146, 147

S
Schachenmayr, Alkuin, 93
Schelling, F. W. J., 114, 115
Schillebeeckx, Edward, 86, 89, 96
Scholarly Record, 1
    corruption of, 16, 24, 43
Self-plagiarism, *see* Text recycling
Semiotica, 69
Semiotics, 69, 70
Semiotic Society of America, 69
Serbia, 129, 132, 133, 139, 140, 145
Shake-and-paste Plagiarism, *see* Plagiarism, Shake-and-paste
Springer, 104, 123
Stanford Encyclopedia of Philosophy, The, 112
Steinke, Hubert, 110, 111
Stemmer, Peter, 16–25, 28, 32, 34

Structural plagiarism, *see* Plagiarism, Structural
Studies in Communication Science, 15, 25, 45
Studies in History and Philosophy of Biological and Biomedical Sciences, 112, 115, 123
Suárez, Francisco, 64
Sunderland, Mary, 105, 106, 108, 123, 124

T
Taylor & Francis, 136, 146
Text-matching software, 2, 13, 37, 40, 46, 152
Text recycling, 5, 8, 45, 51, 52, 63, 68, 70, 72, 134, 135, 146
Theology, 1, 5, 75–84, 91, 92, 94, 97–99, 128
Theology, plagiarism in, 76, 80
Tindale, Christopher W., 38, 42, 43
Toepfer, Georg, 104
Translation Plagiarism, *see* Plagiarism, Translation
Translators, 15, 120
Trembley, Abraham, 105, 106, 111

Tulk, Alfred, 118–120

V
van Eemeren, Frans H., 38, 43, 44
Verbatim plagiarism, *see* Plagiarism, verbatim
Vroniplag Wiki, 153

W
Weber-Wulff, Debora, 2, 6, 13, 14, 33, 41, 94, 109, 152
Western Balkans, 128–133, 140, 141, 145
Whistleblowers
  importance of, 151
  reprisals against, 47
Wieland, Wolfgang, 67–72
Wiley, 54, 58
Wittgenstein, Ludwig, 29
Wolff, Caspar Friedrich, 107–109, 111

Z
Zammito, John, 104
Zimmermann, Johann Georg, 111

The manufacturer's authorised representative in the EU is Springer Nature Customer Service Centre GmbH, Europaplatz 3, 69115 Heidelberg, Germany. If you have any concerns regarding our products, please contact ProductSafety@springernature.com

Printed and bound by CPI Group (UK) Ltd, Croydon, CR0 4YY

25/03/2026

02078174-0018